Alternative Photographic Processes

A resource manual
for the Artist, Photographer,
Craftsperson.

Alternative Photographic Processes

Kent E. Wade

Morgan & Morgan, Inc.

Morgan & Morgan, Inc.
Publishers
145 Palisade Street
Dobbs Ferry, New York 10522

Manufactured in the
United States of America

International Standard Book
Number 0-87100-136-5

Library of Congress Catalog Card
Number 77-83836

Book design by John O'Mara

Color sections printed by
Morgan Press, Incorporated
Dobbs Ferry, New York 10522

I would like to thank the following individuals who so generously contributed visuals, shared their expertise, and/or gave precious time to this project:

Steve Amen, *Portland, Oregon* • Barbara Astman, *Toronto, Ontario* • Martha Forster Banyas, *Portland, Oregon* • John Barna, *Portland, Oregon* • Charles Bigelow, *Portland, Oregon* • Dennis Bookstaber, *Montclair, New Jersey* • David Brown, *Portland, Oregon* • Larry Bullis, *Seattle, Washington* • Fredrich Cantor, *New York, New York* • Ann Chase, *Glencoe, Illinois* • Steven J. Cromwell, *Kansas City, Missouri* • Jean Drews, *Portland, Oregon* • Judy Durick, *Fredonia, New York* • Robert Embrey, *Bellingham, Washington* • Joan Fibiger, *Beaverton, Oregon* • Ford Gilbreath, *Olympia, Washington* • Karen Glaser, *Pittsburgh, Pennsylvania* • Gail Griggs, *Portland, Oregon* • Erik Gronborg, *Solana Beach, California* • Marilyn Higginson, *Portland, Oregon* • Suda House, *Los Angeles, California* • Catherine Jansen, *Wyncote, Pennsylvania* • Christopher P. James, *Greenfield, Massachusetts* • George Johanson, *Portland, Oregon* • Mary Ann Johns, *Salem, Oregon* • Colleen Kenyon, *West Hurley, New York* • Kathleen Kenyon, *Hollywood, California* • Richard A. Kyle, *Upper Montclair, New Jersey* • Ellen Land-Weber, *Arcata, California* • William Larson, *Wyncote, Pennsylvania* • Les Lawrence, *El Cajon, California* • Judy Lehner, *Albany, California* • Jackie Leventhal, *Berkeley, California* • Jill Lynne, *New York, New York* • Paul Marioni, *San Francisco, California* • June Marsh, *Olympia, Washington* • Gehardt Meng, *Portland, Oregon* • Bob Miller, *Portland, Oregon* • Paul Miller, *Portland, Oregon* • Eleanor Moty, *Madison, Wisconsin* • Chris Nelson, *Portland, Oregon* • Bea Nettles, *Rochester, New York* • Roger Ostrom, *Aloha, Oregon* • Peter Pfersick, *Oakland, California* • Richard Posner, *Culver City, California* • Vince Redman, *Portland, Oregon* • Jim Sahlstrand, *Ellensberg, Washington* • Naomi Savage, *Princeton, New Jersey* • Gail Schoelz, *Portland, Oregon* • Diane Sheehan, *West Lafayette, Indiana* • Lori Solondz, *Bloomfield Hills, Michigan* • Gail Skoff, *Berkeley, California* • Stanley R. Smith, *Bellingham, Washington* • Jugo de Vegetales, *Chalfont, Pennsylvania* • David Wolaver, *Portland, Oregon*

I would especially like to thank:

Sherrie Wolf, *Portland, Oregon*, for her excellent drawings;
John O'Mara, *New York, New York*, for the book design;
Elizabeth Burpee, *New York, New York*, for her thoroughness in proofing the manuscript and many worthwhile suggestions;
Dr. Grant Haist, *Rochester, New York*, for his useful comments and list of chemicals and their safe handling procedure;
Liliane De Cock-Morgan and Doug Morgan for their thoughtful advice and personable, professional approach;

And my two best friends: my daughter, Aerin, who frequently offered her assistance and support; and my lover, Susan, for her time, energy, and encouragement to the completion of this book.

Contents

Today's arts and crafts use an astonishing number of hazardous chemicals. For many of you, this will be the first experience with such chemicals. If handled with respect they can be creative, useful tools. If you stick with the basic procedures and precautions outlined below, potential hazards will be lowered and the creative possibilities will be open to you.

1. Keep a written record of all chemicals on hand. Be familiar with their inherent dangers. When you purchase a new chemical find out if there is any possibility of danger in mixing it (intentionally or accidentally) with any chemical you already have. Add this information to your written inventory.

2. When purchasing chemicals, tell the chemist how you will use them and ask about safe handling procedures, including storage (type of container, temperature, etc.). Write the information down in your inventory notebook. When in doubt contact the manufacturer or supplier.

3. Know the antidote for any chemical in case it is accidentally spilled on yourself or taken internally by swallowing or breathing. Note them in your inventory notebook. Have the proper neutralizing agents nearby, together with phone numbers for emergency treatment. (See *Chemicals and Safe Handling Procedures for the Arts and Crafts* in the appendix.)

4. A fire extinguisher is a must for your work area. Be sure it is effective for both chemical and electrical fires.

5. Always use hazardous chemicals outdoors or in a well-ventilated space, preferably under a hood with an air-withdrawal fan. The homemade device illustrated in Chapter 1 could be modified easily to suit individual needs.

6. Always wear the right protection for eyes and hands, as well as any other protective clothing recommended (safety glasses, rubber gloves, plastic apron, etc.).

7. Pour acids into water, never water into acids, and do it slowly. Never pour or mix any chemical at eye level, where a splash could be harmful.

8. Lock up all acids and chemicals and store them—and the keys—away from children, friends, and all who might be unaware of their potential hazard. Be sure children and uninitiated adults stay away while chemical or electrical work is being done. If you must leave temporarily, put someone in charge or lock up carefully.

9. Learn the right way to dispose of each chemical you use: how to neutralize it, what your region's code requires, where and how to dump, etc.

10. Think "safety" at all times and you will be able to use these powerful materials beneficially and safely.

'A Loaf of bread,'
the Walrus said,
'Is what we chiefly need:
Pepper and vinegar besides
Are very good indeed—
Now, if you're ready,
Oysters dear,
We can begin to feed.'

Lewis Carroll, photographer

Introduction

This is a manual of resources—of processes, techniques, and possibilities. It offers "how-to" information for creating images on a variety of surfaces and easy methods for reproducing photographs or drawings used in many printmaking applications. Many of the light-sensitive systems described have been adapted from industrial technology. Others originate in old-time photographic processes, revived and modified for the materials and applications of today. The information has been researched extensively and a useful bibliography and a source list for supplies are provided.

The book has been designed for use as a reference handbook, as a teaching aid for a photography/crafts workshop, or as an advanced guide toward photographic experimentation. There is enough information so that anyone with a rudimentary knowledge of photography, a simple camera, and access to a basic darkroom can make the positives or negatives required for any of the processes discussed.

Information is presented in a functional manner and can be tailored to each reader's requirements. Often the techniques employed for applying images to different materials are similar in methods. Where methods are alike, rather than repeating information a cross-reference is given and only specific differences in procedure are noted. For example, photoetching metal, thoroughly discussed in Chapter 1, provides the basic procedure for photoetching other surfaces as well. Thus, anyone interested in photoetching glass should also read the section on photoetching metal.

The book begins with descriptions of application and processing light-sensitive substances on nonpaper surfaces: Photoetching, enameling, and electroprocessing of metal surfaces; decal fabrication and kiln-fired emulsions for ceramics applications; photosensitizers and dyes for fabrics are discussed. This section is followed by a chapter on altering black-and-white photographs, including such topics as the creative use of bleaching compounds, hand-tinting and toning prints, and selective dye-transfer techniques. Next, nonsilver processes, some predating the silver print are re-examined. These processes not only help to give an historical perspective to photographic development, but also are capable of rendering extraordinary visual delights. Several pages are devoted to the making of high-contrast transparencies, as well as to the construction of images from tonal and/or color separations. In addition, nonphotographic ways to make transparencies from drawings or "found" imagery are presented.

Since many processes rely on photosilkscreening techniques, a chapter is devoted to photostencil preparation and the rudiments of screen printing, including how to assemble a frame and make a vacuum table. The final chapter deals with other imaging systems and introduces some of the new horizons and directions of photography beyond the scope of this book.

Throughout the book many products are mentioned by name. This was done simply because of the success obtained using them, in either my own work or the work of others. There are certainly comparable products on the market that should yield similar results; so be encouraged to experiment.

There are many ways to enhance imagery and/or designs reproduced photographically on alternative surfaces. For example, there are applications for intaglio and silkscreen printing, jewelry, sculpture, stained glass, fiberwork, enameling, and ceramics. Rather than detail each of these subjects extensively, the techniques, materials, and equipment necessary for imaging such surfaces have been emphasized, and the bibliography points out where to go for more advanced information on each subject.

The technology of photography is vast, complex, and continually being refined. An on-going investigation of alternative photographic processes is a means of keeping abreast of these changes and gaining control of the medium. Also the subtleties of how each process can render an image unique will become evident through experimentation. Thus the techniques presented should not be approached as ends in themselves, but as resource tools and creative stimuli with which to strengthen and shape one's seeing. Perhaps, from this view, vision can be communicated more effectively.

This manual was written from the perspective of a photographer in order to share information and to encourage others to share their findings. I hope you have fun with this book and that it becomes a useful tool for expanding visual expression.

Kent E. Wade
Portland, Oregon
February 1978

Photoetching Technique on Metal

Photoetching is a process whereby an image can be transferred to a surface and permanently integrated with it by photographic and chemical means. An object of metal (or glass, ceramic, or some other substrate) is coated with a light-sensitive, acid-resistant compound or film, called a "photoresist." The coated metal surface is then sandwiched with a transparent image and exposed to an actinic light source, such as unfiltered ultraviolet. Where the light passes through the clear areas of the transparency, the photoresist (that is, the coating of chemical compound) is hardened. The resist areas that were masked from the light by the solid (black) areas in the transparency are removed during development. Once the development is completed, a photographic image remains on the metal in the form of an acid-resistant stencil. Acid is then used to eat away (or "etch") the surface of the metal where it is not protected by the resist. The image becomes an integral part of the metal surface in either incised intaglio or raised relief, depending upon the desired effect. This process has been used extensively in the printing industry. Today, photoresists are used to make very precise electronic circuit boards and other small parts photofabricated for industrial purposes.

With this basic understanding, let us examine the use of photoetching techniques by the photographer-artist. There are essentially two ways to approach photoetching: viewing it as a phototransfer process, or using the photoengraved metal as an *objet d'art*. Both alternatives are exciting and both offer sophisticated means of image presentation. If you are interested in the transfer of photographic impressions, then you might consider inking the metal in multiple colors and transferring the image to paper. This is primarily a printmaker's technique. Other alternatives might include the creation of brass-rubbing effects, imprints on leather surfaces, etc. A photo impression can be made on almost any surface, utilizing an etching press and tremendous pressure to assist in the transfer of more subtle tones. On the other hand, using the second approach, the metal itself becomes a visual delight when inked, enameled, electroplated, etc., and mounted for display. Here the metal might also be integrated into other works, such as jewelry, sculpture, etc. Regardless of one's intent, however, the equipment and basic procedure necessary for photoetching the image are the same. There are, of course, some minor variations, depending on the desired end result, and these will be discussed.

THE PHOTORESIST: TYPES AND SOURCES

There are many different types of photoresist materials available:

- a. presensitized photoengraving plates;
- b. dry photosensitive resist films, which adhere easily to metal surfaces;
- c. silkscreened resists not sensitive to light that can be employed where duplicate images are desired (this would not be a very practical method for one-of-a-kind projects);
- d. liquid water-based resists, which do away with the need for special developers;
- e. liquid solvent-based resists, which require a special developer.

The last-named resists (group e.) are the most readily available. They are the least expensive, exhibit the best keeping qualities, and are the easiest to use. In addition they are compatible with the many applications of the photographer-artist. Group e. resists are good, then, for one-of-a-kind productions. Group c. resists would be useful for multiple pieces. Photosilkscreening technique itself will be covered in a separate chapter (Chapter 8).

Each resist uses some kind of developer, although some (as in group d.) need only water for this function. Most solvent-based resists (group e.) have a thinner, in case the resist is packaged too thick for spraying or dipping application. The resists are nearly colorless, and so a dye is usually available in order to make the

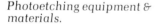

Photoetching equipment & materials.

image more visible after development. Dyes, however, are not normally recommended when maximum chemical resistance is required. All that is usually necessary for photoetching is the photosensitive resist and its companion developer.

Most photoresists can be used when shallow-etching a variety of different metals, including some glass and ceramic surfaces. It is always advisable, however, to use the resist recommended for a particular metal or application. Below are listed several photoresists, together with general comments on their use.

A. A Homemade Photoresist Recipe

If one would enjoy the challenge of making a photoresist at home, then try concocting a mixture of egg white, glue, and ammonium dichromate. The enameline process employed these basic ingredients for etching metal before commercial photoresists were available. For use on either zinc or copper, try the following light-sensitive compound:

- 1. 45ml LePage's fish glue or photoengraving glue, mixed into 30ml distilled water;
- 2. 30ml grade AA fresh egg white (albumin), beaten well with 30ml distilled water (there is approximately 30ml of white to one egg);
- 3. 6 grams ammonium dichromate, dissolved thoroughly in 30ml distilled water;
- 4. 30ml distilled water.

The egg white is prepared first and then slowly mixed into the fish glue. (Filter this colloidal base, if you desire.) Under low light, and just prior to use, add the ammonium dichromate solution to make the colloid sensitive to light. Next add the remaining 30ml of distilled water. More or less of the water may be used, however, according to whether you want a thinner or a thicker resist emulsion. More or less dichromate will increase or decrease the sensitivity of the emulsion.

Apply the mixture to the metal by brush or by pouring a small puddle into the middle of the plate and working it outward to the sides in a thin, even coat. Dry the plate in a dark room, using a hair dryer. Coat the plate a second time, if necessary.

Place a transparency of the image, the same size as you desire the image to appear, on the coated plate and expose to UV light. The areas of the photosensitive coating corresponding to the black in the transparency (and thus hidden from light during exposure) will wash out in a development bath of lukewarm water. It is these "openbite" areas that will later be etched away by a ferric chloride solution because they have no chemical-resistant covering. The areas corresponding to the clear areas of the transparency have been hardened by the light, however, and will resist the bite of the chemical. After development, heat the metal on a hot plate, from underneath, to harden the resist still further. When the resist starts to turn deep brown, but before it turns black, remove the plate from the heat and let it cool. The plate can now be etched, face down.

There are several other dichromated colloids that should make a satisfactory resist for shallow-etching most metals. Experiment with PVC compounds, gum-arabic-based solutions, or dry gelatin-coated

Eleanor Moty, "Portrait Hand Mirror." Image is photo-electroplated in silver rather than photoetched. Object is sterling silver with rosewood handle.

tissue, sensitizing with ammonium dichromate and etching the image with ferric chloride.

Homemade photoresists cannot be expected to perform with the accuracy and resolving ability possible with commercially made products. For example, group e., resin-based resists, are limited only by the quality of the transparency and the exposing technique. One can expect, however, to obtain useful results. For a more comprehensive discussion of colloidal compounds, refer to nonsilver processes in Chapter 7.

B. Kodak Photoresists

1. Kodak Photo Resist (KPR) and KPR Developer. KPR (group e.) is suitable for use in shallow-etching copper and zinc. It is readily available, is packaged in liquid form, and may be either sprayed, poured, or used in dip-coating operations. The dye and the thinner are not necessary for general work. One quart will cover approximately 116 square meters (150 sq. ft.) of surface area.

2. Kodak Thin Film Resist (KTFR). KTFR (group e.) replaces Kodak Metal Etch Resist (KMER), which is sometimes referred to in the literature. It is designed for use with deep-etching of practically any metal, other than copper or copper alloys. KTFR, as packaged, is very thick. Thin it with KTFR thinner in the ratio of approximately one part resist to two parts thinner. (Exposures using KTFR are longer than those suggested for KMER.)

3. Kodak Ortho Resist (KOR). This resist is similar to KPR, except that KOR affords shorter exposures and can be exposed by tungsten rather than ultraviolet light. This introduces the possibility of projecting images onto large or curved metal surfaces. (Normally one is limited to contact images the size of the transparency employed.) KOR does need a very concentrated light source for best results, however. Therefore, as a minimum requirement a 500-watt halogen bulb is suggested for projecting black-and-white slides. It is also feasible to expose resists that are sensitive only to ultraviolet light by means of special equipment, such as a xenon projection system. These projectors are very expensive, however, and experimentation will be necessary.

KOR is one of the more costly resists. You will need both the resist and its companion developer for photoetching.

4. Kodak Photo Resist, Type 3 (KPR₃). Try KPR₃ (group e.) with KOR developer and thinner for electroplating copper surfaces. Mix the developer and thinner about 1:1.

5. Kodak Autopositive Resist, Type 3 (KAR₃). KAR₃ is another good solvent-based resist (group e.) for electroplating. It has its own developer (a concentrate) called Kodak Autopositive Resist Developer. It must be diluted prior to use, one part developer to one part water.

Unlike the other photoresists, KAR₃ creates a positive resist image from a positive transparency. In addition, this resist can be re-exposed after development and partial etching of the metal. All processing, however, must be carried out under safelight conditions. To remove the resist, either use acetone or re-expose the resist to a UV light and redevelop it.

Kodak photoresists and companion products are packaged in

quarts and gallons and can be obtained from any Kodak graphic arts dealer.

C. G.C. Electronics Etch Resist Sensitizer

This solvent-based resist (group e.) is packaged in a spray can. Dye has been added for increased image visibility. Unlike resists packaged for industrial needs, G.C. sensitizer can be purchased in small (three-ounce) cans. This is a suitable quantity for photoetching ten or so small objects. G.C. Electronics P.C. Board Developing Solution, a necessary companion product to the resist, is also packaged in smaller-than-usual containers.

Copper, brass, aluminum, zinc, and glass have been shallow-etched successfully using G.C. resist and developer. The resist has even been used with some success for electroplating. G.C. resist and comparable brands, such as Archer's Etch Resist Sensitizer, are available through radio supply houses, packaged for amateur circuit-board construction. But it is not necessary to purchase an entire kit. Only the sensitizer and developer are required.

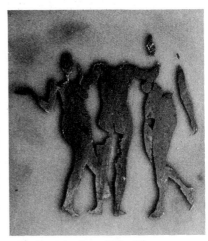

Charles Bigelow, "The Three Graces." A photoetched copper circuit board.

D. Dynachem Resists

Dynachem CMR 5000 Chemical Milling Photo Resist is a useful solvent-based resist (group e.) for deep-etching most metal, glass, and ceramic surfaces. It appears to be similar to Kodak KTFR, except that it can also be used on copper and copper alloys. In addition, it has been used successfully for electroplating. You will need to buy the companion developer. A thinner can also be purchased, in case the solution thickens from evaporation. Resist, thinner, and developer are packaged in gallon sizes.

E. Norland Products

Norland Products, Inc., manufactures water-soluble resists and film-resist systems for industry. One such product is Norland Photo Resist 6 (NPR6), a water-based resist (group d.) with good acid-resistant qualities useful on all metals when ferric chloride is employed as the etchant. This product, however, has a shelf life of only six months.

F. Micro-Metal Presensitized Plates

Presensitized zinc or copper plates (group a.) are mainly used by printers and photoengravers. Here, one has limited flexibility and choice of surface. (Environmentalists are now discouraging the use of zinc plates because they pollute the water system. Thus, many printers are switching to precoated magnesium plates.) These plates come presensitized and are developed in trichloroethylene after being exposed. This chemical is very volatile and should be used only with a glass developing tray in a well-ventilated area. Development times are two to three minutes. The image in resist form can be touched up after development. Prior to etching, the plate should be warmed (this is called the "prebake"). Etch in a bath of nitric acid, diluted one part acid to eight parts water. If you want, you can etch completely through a presensitized plate.

G. Silkscreened Resists

Many different resists on the market are packaged for photosilk-screen application (group c.). These resists, unlike other types, are not sensitive to light. Thus they lend themselves to multiple reproductions. In order to use them, a photostencil must first be made on the silkscreen. Once the stencil has been prepared, the resist can be screened and dried and the image etched with acid. For products and technical information, contact your local silkscreen supplier, the Dynachem Corporation (see D. above), or Atlas Screen Printing Supply, which markets Wornow Printed Circuit Materials. Inquire about Wornow 145-33-P, Etching and Plating Resist.

One can see that there are many different brands and types of photoresists available. Bear in mind that almost all share a common set of methods for applying and developing the resist, as well as for etching the image.

SELECTING THE METAL

Almost any metal can be etched, including copper, zinc, aluminum, brass, and silver. Zinc is relatively inexpensive, compared with copper or brass, and thus it is a good metal on which to practice technique. Zinc is also useful for making intaglio prints. One should inquire at a printmakers' supply house for zinc plates with an acid-resistant back. Other metals must be coated on the back and edges with either a photoresist or asphaltum, prior to etching.

Most metals can be obtained from metalworks. In addition, some jewelry supply houses carry small sheets of copper, brass, or silver. Consideration should also be given to "found" objects or copper-coated circuit-board material as other metallic surfaces on which to work. Finally, there are the plating companies, who can offer mul-tilayered sheets of metal.

Copper is a hard metal that will hold very fine detail. Zinc is softer in molecular structure and tends to give a more ragged etched line. Silver is a fine metal for special projects, but it is expensive. It is possible to save some of the cost by electroplating copper with silver or a silver-colored metal, such as nickel. These will give a similar visual effect. Generally, one does not need a very thick gauge of metal for etching. A 22-gauge is sufficient. If you are planning to enamel the image, a thicker metal (12-gauge to 18-gauge) is recommended.

Some suppliers will cut the metal to size; or you can do it yourself. A plate can be cut into almost any design with a hand-held jigsaw, provided you use the appropriate blade for sawing metal, a dab of oil on the cutting line, and a great deal of patience. If a piece is going to be electroplated, leave an extra margin of metal. This part (which should be removed later) can be drilled and used as the electrical connection point during electroplating. The holes in it can also be useful as points by which to suspend the plate during the photoresist processing steps.

CLEANING THE METAL BEFORE APPLYING RESIST

Even if the metal is not scratched, it should be polished and finished prior to application of the resist, as well as in preparation for final display. Buy sandpaper, such as 3M Carborundum Waterproof Metal Sandpaper. Purchase it in three sizes: coarse #240, medium #400, and fine #600. Sand the metal, beginning with the coarsest grade paper; change to the medium, and then to the fine grade for finish-

Cutting up pieces of zinc for photoetching.

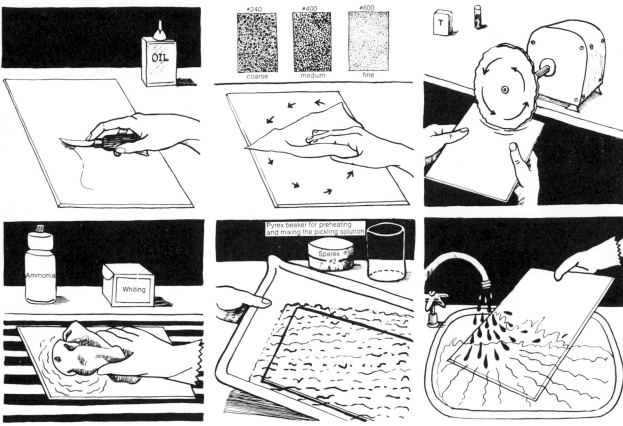

#240 coarse #400 medium #600 fine

Pyrex beaker for preheating and mixing the pickling solution

Sparex ® #2

Ammonia

Whiting

OIL

T

Stages of cleaning the metal: (left to right from top). 1. Scrape and burnish the metal surface. 2. & 3. Sand and buff the plate. 4. & 5. Degrease and pickle the plate. 6. Final water rinse of the plate prior to resist application.

ing. Prior to sanding, where deep scratches are evident a curved burnisher and a scraper can be helpful. First scrape the metal, using even strokes, until the scratched area is relatively smooth. Next, burnish the area with all-purpose oil until the surface indentation has been removed. Finally the plate can be sanded as outlined above. Once the scratches are gone, buff the metal with jeweler's tripoli on a buffing wheel. After washing thoroughly with ammonia and hot water, follow up with a polishing of jeweler's rouge. Although it isn't absolutely necessary to scrape, burnish, sand, buff, or polish, this exercise helps to create the optimum surface for resist application.

Before applying the photoresist, you must degrease the plate. This is accomplished by rubbing it with whiting (ground-up white chalk) and household ammonia. Pumice powder can also be used. Next, rinse the plate in water. When water runs over the plate without beading up, the surface is degreased and clean. It is very important that the plate be clean if the resist is to adhere. As a precaution, wear cotton gloves, so as not to get finger marks or skin oil on the plate. Cleaning is best done just before you apply the resist, so as to keep the plate free of dust and to minimize oxidation (tarnish). It cannot be stressed enough that, regardless of the cleaning method employed, a clean plate will save hours of grief.

Special Preparation for Metal Surfaces
It is sometimes advisable to pickle the metal, after degreasing it, in a

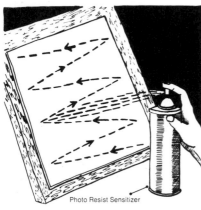

Photo Resist Sensitizer

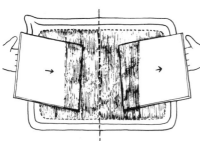

Three methods of resist application: (top to bottom). 1. Pouring the resist. 2. The spraying method. 3. Dipping the plate in resist. (Use one continuous movement.)

weak acid bath to remove heavy oxide deposits that might cause the resist to lift off. A weak nitric acid or hydrochloric acid solution can be used with most metals; however, a safer pickling bath can be made from Sparex No. 2, a granular acid compound. This is used by jewelers for pickling nonferrous metals, such as copper and silver, and is classified as a nonhazardous material. Make a working solution by thoroughly dissolving 300ml (10 oz.) of Sparex into 960ml of water in a glass container. Before using, heat this solution to 60° C (140° F). Sparex is available from jewelry supply houses.

When a resist-coated metal is being etched, the acid tends to bite at an angle rather than straight down, undercutting the resist coating. Treatment with a conversion coating, which increases resist adhesion and redirects the acid's bite, helps to reduce this problem. Although it is not necessary to use a conversion coating, a sharper image can be achieved and the finer detail retained, especially when one is deep-etching or etching halftone images. Conversion coatings are available for most metals, and formulas may be obtained from Eastman Kodak. To make a conversion coating for zinc, use a 2% solution of phosphoric acid. It is bottled in a 85% solution; therefore, mix 30ml of acid with 600ml of water to obtain a suitable working solution. *(Always pour acid into water, never water into acid, and work slowly. Wear gloves and goggles especially when working with undiluted acids.)* Immerse the plate for approximately two minutes at room temperature, with mild agitation, then rinse and dry it. The metal is now ready to be coated with photoresist.

APPLYING THE PHOTORESIST

There are numerous ways to apply a liquid photoresist. The easiest involve spraying the resist, pouring the resist, or dipping the object.

If a spray system is employed, first tape the metal to a board, using double-sided adhesive tape. (Coat the back of the plate first.) Spraying can be done with a spray gun and a properly diluted resist, or with a resist packaged in a spray container. Spray from approximately 20cm (8 in.) away, starting with the bottom of the plate and, using a back-and-forth motion, working upward to the top. A reasonably consistent coat is desired on the side that will be etched. Generally, the finer the detail in the image, the thinner this coat should be. Bolder images can tolerate a thicker coating. An uneven coating can affect exposures to some degree, as well as lead to premature resist-failure because one area is denser than another. This problem can be minimized by respraying with the plate turned 90°. Spray a thin yet generous coat, so that the resist runs slightly and creates a continuous film over the entire plate. Spraying in rapid, jerky motions will cause spottiness and unevenness.

Experience quickly teaches the finer points in applying the resist. If any dust particles are seen in the resist coating, or if it is extremely spotty or uneven, then remove the coating with lacquer thinner *(wear gloves)*. The plate must then be recleaned with household ammonia and whiting before another coating of resist is applied.

Do not be concerned about the quality of the resist coating at the edges of the plate. More than likely the image will be situated in the center area. The edges can always be touched up with asphaltum prior to etching.

Dip-coating a plate is also easy. Here, the metal is immersed in a pan of resist. For optimum coating thickness, try to maintain an uninterrupted withdrawal rate of approximately 15-to-23cm (5-to-9 in.) per minute. In addition, be alert to a possible wedging effect caused by the way a liquid resist runs off a plate. If after drying the resist coat appears to be thin at one end, then redip the plate from the opposite direction to even out the coating's thickness.

Another method of application requires that a small puddle of resist be placed in the middle of the plate. Work it in a circular motion to the outer edges until the thickness of the coating appears to be even. A piece of cardstock can be helpful in working the resist around the plate.

Dip-coating is recommended for metals other than zinc because an acid-resistant coating can be created on both the front and the back of the plate in one easy step. With spray or pouring technique, the back should be coated separately and dried before the front can be done. Alternatively, the back can be coated after development of the resist image with asphaltum or stop-out varnish (from a print-making supply house).

Choose the method of resist-application that comes the easiest. Again, remember that the objective is a reasonably even coating free of dust and bubbles. When applying the resist, do so in a dust-free area and work in subdued light. A low-watt, yellow insect-repellent lamp is fine. A red safelight is even better. As the resist dries, it does become sensitive to ultraviolet waves, so keep the unexposed plate in a dark place.

Photosilkscreening Resists Onto Metal

Photosilkscreening is a time-saving technique for duplicating an image, especially when making ten or more identical plates. The technique itself involves the photographic formation of an image (a stencil) on the silkscreen, through which a resist can be forced (using a rubber blade, or "squeegee") to produce the image on the metal. Images screened in resist mediums, however, are never as sharp as images formed directly on the plate by photoprojection or by contact-printing. This difference, though, may be inconsequential and not even apparent to the untrained eye.

A resist can be made at home using 100% asphaltum or a mixture of asphaltum (75%) and beeswax (25%). Thin either resist with mineral spirits until just diluted enough to pass through the screen. There are also commercially made resists, available from silkscreen dealers. These resists must be prebaked at approximately 83°C (200° F) to remove excess solvents before etching. A stainless steel or a monofilament polyester screen is recommended, and standard photosilkscreening methods can be used (see Chapter 9).

PREBAKE DRYING PROCEDURE

The function of the prebake cycle is to remove any solvents that remain in the photoresist coating prior to exposing it to ultraviolet rays. This will insure optimum resist adhesion and sharper image formation.

A light proof drying box: any size to meet your needs.

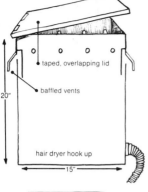

taped, overlapping lid

baffled vents

20"

hair dryer hook up

15"

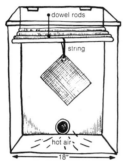

dowel rods

string

hot air

18"

If you coat several plates at one time, make a small drying chamber out of a lightproof cardboard box. Secure some dowels through the top portion of the box, so that the coated plates can be suspended while drying. Use a metal plate larger than the image desired. The excess plate material can be drilled with holes on all four corners and strings attached. The plate can then be suspended in the drying box as well as later on, upside down, for etching in ferric chloride. In the bottom of the box make a small hole to connect a hairdryer hose. At the top of the box prepare a lightproof vent, or baffled opening, to permit air circulation.

Plates can also be coated one at a time and a hair dryer with medium heat used to warm, or "prebake," the resist layer. The objective is to evaporate excess solvents from the plate; otherwise, exposures might be affected. Ideally, an infrared drying system should be employed. This system allows the resist to dry from the plate surface outward, avoiding the possible entrapment of solvents.

Each resist has temperature limitations; so be sure to follow the directions for prebake. Generally, too much heat (82° C–180° F— for longer than eight minutes is maximum for most resists) will tend to fog the resist and affect normal exposure and development.

Before prebaking, let the coated plate air-dry for about 15 minutes in subdued light at approximately 20° C (68° F).

The basic rules for the prebake procedure are: Avoid overbaking the plate; dry thoroughly with minimal heat under low light; make sure there is enough ventilation.

EXPOSING THE PHOTORESIST

There are several points that need to be made regarding plate exposure:

1. The resist is affected by the ultraviolet part of the spectrum. Therefore, in order to form an image in the resist layer an ultraviolet source—such as black light, mercury vapor lamp, sun lamp, carbon arc, quartz lamp, or even direct sunlight—must be used. A 500-watt photoflood can also be used, but it is the least effective of the UV sources. For occasional work, an inexpensive sun lamp is suggested; however, as with any of these light sources, be aware of potential eye damage (e.g., pinkeye) and burns.

Generally, exposures should be made at a distance equal to the diagonal of the metal plate. This distance seems to provide even illumination with no visible hot spots. These light sources can generate a great deal of heat, however; so on plates 20x25cm (8x10 in.) or smaller, expose at no less than 38cm (15 in.). If both sides of the plate are coated with photoresist, first expose the back side without any transparency in order to create an acid-resistant barrier. (The resist on the back must be hardened so that it won't wash off during development.) Next expose the image side with a transparency in position, emulsion side down.

2. The length of the exposure is directly affected by the thickness of the photoresist coating, density of the image on the transparency, and intensity of the light source. It is advisable to run a test plate with the resist to establish an average exposure range. Exposures usually vary from three to seven minutes with a sun lamp at approximately 38cm (15 in.) from the surface of the resist. If a transparency has some light gray tones or very fine detail that is important to the image, longer exposures will harden the resist in those areas. Thus, when running the exposure test it is best to use a transparency that has a complete range of tones. Use two-minute intervals for the test. If during normal development of the plate the resist washes off completely, the plate was probably underexposed. If the photoresist has been heavily overexposed, even those areas corresponding to some of the darker areas in the transparency will probably harden.

Normally there is greater latitude for overexposure than for underexposure. Generally, overexposure of halftone highlights or of any fine detail should be construed as one limit, and a general underexposure of the entire plate, the other. You will be working to achieve exposure times within these boundaries.

Each transparency will usually require a slightly different exposure. It is a good idea to keep test strips and review them against the density range of any new transparency about which you may have doubts. It only requires a quick examination to obtain a new approximate exposure time. Kodak markets a test strip called a Kodak Photographic Step Tablet No. 2, which might be helpful in establishing an optimum exposure range for any transparency. It can be calibrated with a densitometer or given an arbitrary number system. Mark the step tablet with opaque ink to correspond with its tonal scale. This density scale will relate directly to the various tones of any transparency utilized.

3. Close contact between the coated metal surface and the transparency is very important. One may use a vacuum-type frame with a built-in UV light source to insure consistently good results. However, a piece of single-strength clear glass can also effectively sandwich the transparency flat to the resist-coated plate. It is possible to obtain very good contact with this simpler, less expensive method. If the transparency appears to lift off the plate at all, one can use C-clamps at the corners to maintain maximum contact. A contact-printing frame may also be utilized. Whatever the system, the transparency should be sandwiched tightly with its emulsion side to the photoresist layer on the metal plate.

4. Chapter 8 is devoted to the alternatives and techniques available for making a transparency. Only a few basic points that relate to transparencies for photoetching will be discussed here.

a. A positive transparency is required to obtain a "positive-reading" image in the etched metal. Images etched in this manner are usually to be placed on display. When a plate is destined for image transfer (e.g., to be inked and have prints pulled from it), it is normally made with a negative transparency, so that the image will "read" positive. A plate made in this manner is usually inked in its recesses (intaglio). A positive-reading image on a plate can also be inked to produce a positive-reading print if a roller is used to ink the image in relief, or woodblock, fashion. Of course, positive and negative transparencies can be combined for photoetching the plate, with a variety of techniques used to ink the plate.

b. Unless a projection system is employed, the transparency must be the same size as the final image desired.

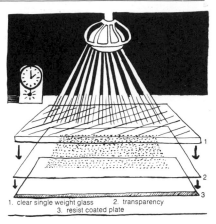

1. clear single weight glass   2. transparency
3. resist coated plate

Plate being exposed.

c. High-contrast film, such as Kodalith, is normally used to make the transparency; however, when working with irregular metal surfaces it may be difficult to make an Estar-based transparency conform to the curves and bends. Instead, try using an acetate-backed high-contrast film. The acetate backing can be softened by immersing the transparency for seven minutes in a solution of acetone and water, mixed 1:1. Place the softened film on the surface to be etched, working out the trapped moisture and air. Allow the film to dry. During drying and exposure of the film, it may be helpful to place the object in a plastic bag under vacuum in order to improve contact between the film and the metal. One might also try Kodalith Transparent Stripping Film 6554 (Type 3). While this film is still wet from its final stage in the development process, peel off the backing and place the film on the resist-coated object, emulsion side down. In this case, make the exposure while the film is still wet and in good contact with the piece. Finally, remove the film and develop the photoresist.

5. A slide projector may be used occasionally for exposing certain resists. A high-contrast positive transparency will be needed. Make a slide, using Kodak Direct Positive Pan Film 5246 to photograph the original artwork. This film is designed for reversal processing in Kodak's reversal-processing kit. Alternatively, Kodak's Panatomic-X may also be used, if processed in the same kit. Once either film has been processed, it can be mounted for projection.

6. Sometimes you might want to etch a round surface. In order to expose the resist, lay the transparency on the surface, emulsion side to photoresist, and tape it down. Then set the object on a record turntable. Place a UV light source perpendicular to it. As the turntable rotates, the resist-coated surface is exposed. Exposure can be calculated by running a test strip, making certain that the turntable is weighted comparably to the weight of the object that is to be exposed. (A heavy object can slow down a sensitive turntable significantly and affect the exposure time.) A narrow slit cut into a piece of cardboard and mounted between the light source and the revolving object affords the best means of controlling the light striking the surface: Only a small portion of the resist will be exposed at any one time, and less light will be able to bleed off along the curves of the piece and complicate the exposure time.

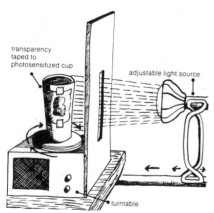

Setup for exposure of round surfaces.

DEVELOPING THE PHOTORESIST

Once the plate has been exposed, it is placed in the appropriate developing solution. Development should be performed shortly after exposure and in subdued light. Almost immediately a faint image will begin to appear. Develop with agitation for no more than two minutes. Overdevelopment tends to soften the hardened resist areas, swelling and lifting the resist off the plate. In order to make sure the unexposed resist is completely removed develop the plate for at least one minute. Keep the plate face up during development to avoid scratches on the surface. Be very careful when handling the plate, as it is easily marred during and immediately after development.

Because of the nature of the developing solutions, *caution is* called for in various ways: use only a glass or aluminum tray • do not allow the developer to get on rubber gloves or plastic containers, as it will soften them • developing solutions are toxic; therefore, provide adequate ventilation and keep the solutions off your hands • review notes on chemicals • do not work in the kitchen • wear goggles, a respirator with the appropriate filter pack, and developer-resistant gloves (check with safety equipment supplier).

Although xylene is occasionally employed as a substitute developer with some resists, it is best to stay with the manufacturer's recommendations. Xylene fumes, like those of most other resist developers, *should not be inhaled.*

Once the plate has been developed, transfer it to a pan of flowing water and agitate it for about one minute. Next, tilt the plate upright and give it a fine spray of cool water. Use a plant-misting device. The developer will not mix with the water, but do not be concerned: The mist application insures that all the unexposed resist is removed prior to the postbake treatment (described later).

After development, room lights may be turned on for inspection. Surface-dry the plate by blowing excess moisture off with warm air from a hair dryer. Next use a magnifying glass to examine the faint image. Places where the resist is "pinholed" will be obvious. The edges of the image will probably look sloppy. These can all be touched up, but make certain that the image looks complete and not just half-exposed or half-developed. Final examination of the image can sometimes be difficult; therefore, immerse the piece in a ferric chloride solution for about 30 seconds. Black residue from this quick dip indicates where the plate will etch. Rinse the plate in water and do any spotting necessary with asphaltum resist (more about this later).

Troubleshooting Development Problems

If the resist lifts completely off the plate, you can be reasonably certain that the resist was underexposed or not completely dry at the time of exposure or that the metal surface was not properly cleaned. (Fingerprints seem to be the worst offenders.)

Sometimes the resist may swell up slightly off the surface of the plate. This problem is usually due to underexposure, too thick a resist coating, or, as mentioned earlier, too much time in the developing solution.

Another potential problem comes from contaminated developer. The image will appear to be developed, but in fact the unexposed areas will not be developed completely free of resist. Thus, after the postbake, the etchant will not be able to react with the metal. Contamination is often due to using the developing solution too many times. The developer tends to lose its working properties because of a build-up of resist globules that wash off the plate during development. It is imperative, therefore, to change the developer periodically to avoid this problem.

An image may not develop out completely because of overexposure. When the resist hardens in the finely detailed or halftone highlight areas, one is clued to an overexposure problem. In addition, excessive heat in the prebake cycle or pre-exposure to extraneous UV light will also "fog" or harden a resist coating.

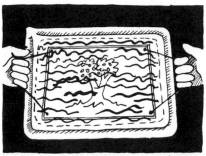

Metal plate being developed in a glass tray and hand misting plate to remove residual resist.

Postbaking a plate.

Photoresist touch-up with Universal Etching Ground.

A simple rule to help avoid problems with resist application and development:

Careful, thoughtful work habits are one's best guarantee to keep the task at hand completely trouble-free.

Postbaking the Image

Baking the plate after development hardens the photoresist considerably, increasing its resistance to the acid as well as its durability in the electroplating process. Bake the plate at approximately 93°-122° C (200°-250° F) for about ten minutes in an ordinary oven. Remember that poisonous solvents evaporate from the plate; so it may prove dangerous to use the same oven in which food is cooked. Postbaking is not really necessary for shallow-etching, but it is recommended to thwart image liftoff. The postbake removes any solvents trapped inside the resist after the surface has hardened.

IMAGE TOUCHUP AND METAL PROTECTION PRIOR TO ETCHING

Use asphaltum, or stop-out varnish, which dries faster, to cover all metallic surfaces that are not to be etched. This includes pinholes in the resist coating, the edges of the plate, the back of the plate, undesirable areas of the image, and so on. Any fine, small brush can be used. Once the touch-up medium has dried thoroughly, inside as well as on the surface, the plate can be etched. When necessary, asphaltum can be diluted with turpentine for ease in application, and also for removal of touch-up mistakes, as this solvent is gentle on the photoresist itself.

ETCHING THE METAL: Chemical Solutions and Techniques

"Etching" refers to the removal of metal from the plate by the action of some corrosive agent (acid). The resist used on the plate prevents the acid from "nibbling" away the metal in those areas where the resist has been applied. The acids or etchants used are dangerous; so here are some recommendations for safe handling:

- 1. always wear good rubber gloves and goggles;
- 2. never pour acid at eye level;
- 3. always pour acid into water (not vice versa), and pour it slowly;
- 4. never mix anything with acid unless you first check it out with a chemist (for example, any combination of cyanide and acid will react to form lethal cyanide gas);
- 5. for each acid, check at a chemical supply house as to proper neutralizing agents; buy the appropriate ones and keep them handy in case of accident;
- 6. make sure acids and chemicals are well labeled;
- 7. keep acids and chemicals out of the reach of children, preferably in a locked place with the key out of reach;
- 8. acid fumes should not be inhaled: use acids only in a well-ventilated work area with an exhaust device—or otherwise work outdoors (a respirator with an interchangeable filter pack is a good investment);
- 9. use heavy-duty plastic darkroom trays for all diluted acid baths or mordants;

● 10. never use the same acid bath for different metals, as this can cause contamination of the acid;

● 11. always lower the metal into the bath slowly, yet evenly, to insure that all parts receive a similar amount of etching time.

There are many suitable etching baths, or mordants. Those recommended are listed below.

1. Nitric Acid: Nitric acid is useful for etching a variety of metals. Caution is called for, as this acid does liberate a poisonous gas when it reacts with the metal; however, the bubbles that form during the reaction are useful in indicating where and how rapidly the metal is being etched. These bubbles should be removed periodically because they hinder the etching progress. Lightly brush them off the surface of the plate with a feather.

Collapsible acid etching box with clear plastic cover.

Begin to etch the metal by inserting the plate, image side up, into the solution. Proceed until the first halftone highlight areas start to vanish, or until the acid starts to undercut and cause the finer lines and minute detail to disappear. There are so many uncontrollable variables that experience and continual observation are the only reliable teachers as to how long the plate should be etched. Temperature, for example, affects etching time. Normally, use nitric acid at room temperature, although one can speed up the action by warming the mordant, or etchant in a pan of hot water.

Generally, when using a strong nitric mordant it is best to remove the plate every three minutes, rinse it in water, and inspect it. With a weaker nitric solution, five-minute intervals are sufficient. Sometimes when a piece is being etched deeply, 20 minutes between inspections might be enough. With repeated usage, any acid bath will lose its strength and the entire process will take increasingly longer. When this happens, freshen the solution or change the bath completely. Try the following dilutions, altering the strength as required:

● a. *For zinc:* Mix one part nitric acid to six parts water for normal etching.

● b. *For copper or brass:* Use three parts nitric acid to five parts water. These metals are harder and require a stronger bath.

● c. *For silver:* Use one part nitric acid to eight parts water.

Changing the above dilutions will alter the rate of etch.

2. Dutch Bath for Copper: This formula is designed for etching fine lines in copper. 75ml of hydrochloric acid, 15 grams of potassium chlorate, and 375ml of water are required. First boil the water, then pour it over the potassium chlorate crystals to dissolve them. Next cool the solution to room temperature or 20° C (68° F) and add the acid. Take note that the Dutch bath does not bubble; so the plate must be checked continually.

3. Ferric Chloride: This is one of the safer etches available. Try to purchase the industrial grade prepared for etching copper curcuit boards. This mordant will etch aluminum, brass, and zinc satisfactorily, as well as copper. It will also etch silver, but it may take three-to-five hours to obtain a satisfactory image. If you have to buy ferric chloride in crystal form, invest also in an inexpensive Baumé hydrometer. This device is used to measure the densities of solutions. Make a solution by crushing ferric chloride crystals, then

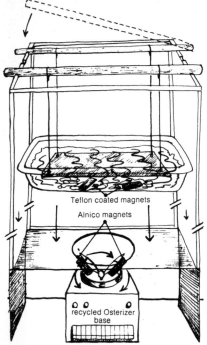

Teflon coated magnets

Alnico magnets

recycled Osterizer base

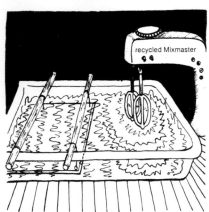

recycled Mixmaster

Agitation systems for etching with ferric chloride.

dissolving the powder in water. When the solution registers in the proper range—36° to 42° Baumé—it is ready to be used as an etchant. Generally, it is the proper consistency when it is syrupy.

To use this etchant, heat it to approximately 38° C (100° F). Heat can dramatically affect the rate of etching; so reheat the solution when the temperature drops to approximately 27° C (80° F). Because ferric chloride does not bubble, it is preferable to etch the plate upside down with constant agitation. Bubbles help to remove the sediment that forms in the lines as the acid bites the metal. Suspending the plate upside down permits the residue to fall freely, insuring that the etchant will act uniformly and give a cleaner edge to the final image.

Absence of bubbles makes it difficult to check the progress of the etch. The solution can be modified to release bubbles, however, by adding 30ml of hydrochloric acid to every 750ml of ferric chloride solution. As the bubbles also help to remove sediment, the plate can be etched image-side up in this modified etch. Constant agitation should still be given, and a feather should be brushed across the plate periodically to make certain all sediment is removed from the non-resist areas.

A halftone image can be etched completely in 20 minutes, while a piece etched selectively may require several hours. Agitating a piece for two hours by hand can be monotonous; therefore, some alternatives are presented here to free you for other tasks during less critical times:

One method involves suspending the plate upside down in the etchant from four strings attached to corner holes drilled in the excess margin of the plate. An old electric kitchen beater (e.g., Mixmaster) is used to agitate the solution at a slow speed. Place a single beater arm in the etchant at one end of the tray, away from the object to avoid scratching it. (Coat the metal beaters with asphaltum or a plastic varnish to protect them from the acid.) If necessary, prop up one end of the etching tray in order to keep the solution level. Check the etching action every 10-to-30 minutes.

One might also try placing Teflon-coated stirring magnets (available from chemical supply houses) in the solution, keeping them in motion by attaching an Alnico magnet to the blades of a discarded Osterizer base. First a wooden frame must be built around the Osterizer, leaving an opening where the Alnico magnet spins. The frame should be designed so that the rotating magnet will be slightly below the plastic tray containing the etchant, yet close enough to influence the magnetic stirrers. Above the tray, an open frame can be constructed to which the strings running from the plate can be secured in order to suspend the plate upside down in the solution.

Ferric chloride is an unusual etchant in that it will blacken the areas and lines where it bites the metal. This by-product of the etching action is useful in two ways: *a.* the black colorant may be left on the plate for decorative purposes; *b.* it can be used to check visually the progress of the etch. It also provides an early warning system of defective areas in the resist: One can actually see whether the plate is being etched only in the nonresist areas and that the image is not being undercut. Monitor progress by removing the plate

from the acid periodically and rinsing off the black residue with water. The depth of the bite and the tone that will be created in printing from the plate can be determined by timing how long the etch solution takes to blacken the image. For example, experience might indicate that once the image has gone deep black four times (removing the color with water each time and returning the plate to the etchant), a shallow line will have been created. If the plate was inked and printed without further etching, it would produce a reasonably light tone. If a darker tone were desired, the plate would have to be etched for a longer period.

Of course at any time during the etching process, should the resist break down or should you desire to control print tones or depth of the bite in selective areas of the image, asphaltum can be applied. If a plate is going to be resensitized, make certain that all ferric chloride residue is removed before applying fresh resist.

REMOVING THE RESIST (Optional)

Photoresist can sometimes be difficult to remove; therefore, it is often left on the object to serve as a protective coating to the metal. If the resist must be removed, then try using lacquer thinner or soaking the plate in the resist's companion developer. It can also be burned off with a torch, or fine sandpaper can be used, followed by a rebuffing of the plate. Turpentine is a good solvent to remove the asphaltum that was added for plate protection.

REWORKING THE PLATE

The fact that a plate can be reworked opens up a new realm of possibilities for image interpretation. Besides, it is seldom that one is satisfied with a "straight" photoetching. Some of the methods and reasons for re-etching a plate are listed below:

1. Aquatint. When a plate is destined to be inked and printed, the etched areas are often as smooth as the unetched areas and will not hold ink. This is particularly evident when the transparency is anything other than a fine-line or a halftone reproduction. Aquatinting is a printmaker's technique for creating tone in the plate.

Without removing the resist, sprinkle powdered rosin through a piece of nylon stocking, evenly covering the surface of the plate. The powder should cover only about half the total surface area. Bare metal should still be visible through the powder. Next, heat the plate to melt the rosin, so that it adheres slightly and can act as a resist. Re-etch the plate in an acid bath, normal dilution. A granular surface will be created on the previously smooth metal areas, improving the plate's ability to hold ink. Aquatinting procedure may need repeating over the same area and/or different areas of the plate before the desired range of tones is attained.

A wide range of tones is possible from re-etching the plate for short periods. Delicate gray tones can be etched in as little as 20 seconds, while rich blacks can be achieved in four minutes. Enamel paint, sprayed from a gun or can, also has been utilized effectively to create an aquatint of unusually fine tonal quality. This alternative is much faster than traditional methods.

Try making original images on ultrasensitive Kodak 2475 Re-

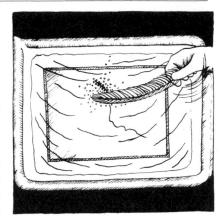

Feather brushing away bubbles on the etching line to improve the etch.

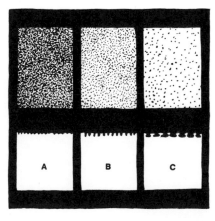

Aquatint rosin on plate: A) Too heavy a rosin coating. B) Perfect 50/50 coverage of rosin. C) Too thin a rosin coating causes undercutting.

Rosin ready to be ground and being dusted on.

cording Film. This film tends to give a very grainy continuous-tone-like image when used to make enlarged high-contrast transparencies. Such images, when photoetched on metal, tend to hold ink surprisingly well. There are still more possibilities worthy of consideration. Try aquatinting the plate prior to photoetching, or experiment with using a texture screen when making transparencies. Printmakers sometimes use nonglare glass, which tends to create a mezzotint effect on the transparency. A piece of sandblasted glass, with the frosted side down on the film during exposure, also gives a very pleasing, grainy quality to an image.

2. Posterization. Sometimes a plate is re-etched several times for a posterlike effect. Each time a high-contrast tonal separation is used representing the highlights, midtone, or shadow areas of the image. In this way the lines of the image are emphasized, giving it added dimension. Re-etching procedure is simple. It involves recleaning the plate, reapplying the photoresist, realigning (or "registering") another transparency over the etched image created by the first, and processing as before.

Kodak KAR₃ is an unusual photoresist in that it can be re-exposed and redeveloped without having to apply additional coatings of resist. All processing must be done, however, under safelight conditions. Although this resist is expensive, it is an excellent choice for multiple etching or plating (see posterization technique in Chapter 8).

3. Relief. After the image has been etched, remove any photoresist from areas surrounding the image. Carefully cover the main elements of the image with asphaltum, leaving unimportant detail and background areas uncovered. Next re-etch the plate in an extra-strong nitric bath designed for deep etching. This nibbling away of nonessential information can strengthen visual impact, as the image will now be isolated and will protrude slightly from the surface of the plate. Relief technique is very useful for plates destined to be put on display.

Universal Etching Ground is not as resistant to a strong acid bath as a heavy asphaltum coating is. Asphaltum, however, normally takes a very long time to dry, about an hour or so. Drying can be expedited by setting a butane torch to the asphaltum and burning out the excess solvent. This should be done in a safe place outdoors, with a fire extinguisher close at hand. A plate can be dried in ten minutes this way. One must be extremely careful of the flames during the burn-out; so double check the area for fire hazards before "torching" a plate. This technique is not recommended for plates larger than 20x25cm (8x10 in.). Set larger plates on an etcher's hot-plate box in front of a hair dryer.

4. Border. A border design can be pleasing on plates to be displayed. Make a transparency of the chosen design and photoetch the border first. Next photoetch the image. Follow up by creating a relief effect around the image by selective retouching and re-etching, as described above.

5. Multiple Imagery. Photoetching need not be limited to single-image presentation. Transparencies can be combined by splicing them together with tape, or several different images can be worked into the plate by retouching and re-etching techniques.

6. Adding Hand-Drawn Designs. Cover the plate with Universal Etching Ground and then scrape a design through the ground. Be sure it reaches down to the metal. Etch the design, clean the plate, and photoetch an image atop the drawing. A drawing on tracing paper can be transferred quickly onto a photoresist, using the thin paper as you would a film transparency.

7. Selective Etching. To hold fine detail or vary the depth of the line, asphaltum can be used to block out selected segments of the image, even during the etching process. For example, the image can be etched, rinsed, retouched with asphaltum, dried, re-etched, re-rinsed, then more asphaltum added, etc., until the desired effect is achieved.

8. Line Images. Sandwich high-contrast positive and negative transparencies of the same image together in order to create a line effect on the plate. This idea may be carried further by re-etching with another positive, of the same image but in a different density.

These are just a few ideas to titillate the imagination. Of course many of these methods can be combined.

You will soon find that the art of photoetching demands a great deal of previsualization. Planning an image and its presentation is more than half the task. The photoetching process itself is easy.

ETCHING MULTILAYERED METALS and THROUGHPLATE ETCHING

It can be rather exciting to etch through a layer of nickel (the same color as silver) and expose a layer of copper, for now color is added to the plate. Electroplating is the method for coating the copper sub-stratum with nickel. It is really very easy, although, if you prefer, plating companies will coat a sheet of metal with three or more layers, so that, when it is etched selectively with acid or with electricity, different contrasting metals are exposed. Advise the electroplating company that a thick coat isn't essential, since you only intend to expose the layer of metal below it. Some knowledge of metals and of "discriminating" acids or electrolytes that only etch or remove one layer of metal at a time, would be useful. More about electroprocesses at the end of this chapter.

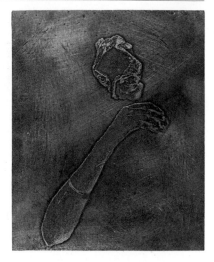

Kent E. Wade, "Untitled." A photoengraving: photoetched brass plate, image was blocked out and background deeply etched to create relief. Plate was then oxidized and lacquered. 4" x 5".

Used lithographic offset plates can be purchased inexpensively for throughplate etching. These thin metal plates are very flexible and easily mounted by gluing to any solid support, such as wood. With the appropriate subject matter, a plate that shows the texture of the wood through the openbite areas could be visually effective. These void areas can also be filled with clear resin (epoxy) to make the surface of the plate flush, yet permit the color of the support to be seen. Other materials (paints, enamel, etc.) could replace the resin affording the artist a limitless choice of color and texture.

REVIEW OF PHOTOETCHING STEPS

1. clean plate **2.** apply photoresist **3.** prebake the resist to rid it of solvents **4.** expose the plate to an ultraviolet light source **5.** develop the resist **6.** postbake the resist to harden it and improve its acid resistance **7.** do touch-up work with asphaltum or Universal Etching Ground **8.** etch with the appropriate acid solution **9.** rework the plate, if desired **10.** remove the resist and asphaltum

PHOTOETCHING MATERIALS CHECKLIST

● metal to be etched ● jigsaw with ferrous and nonferrous blades for cutting plates ● sandpaper: rough, medium, and fine grades for metal work ● tripoli and jeweler's rouge, plus buffing wheel and buffer ● file to bevel edges of a plate ● a burnisher and a scraper ● whiting or a fine grade of pumice powder that won't scratch the metal's surface ● fire extinguisher, for the many highly flammable solvents ● household ammonia ● photosensitive resist and its companion products (developer and thinner) ● spray gun if resist is to be sprayed ● hair dryer to speed up the drying steps ● sun lamp or other ultraviolet light source ● ferric chloride etchant or other appropriate acid bath ● brown glass storage bottles for the acids ● necessary antidotes for each acid, in case of accident ● acid-resistant rubber gloves, developer-resistant gloves, plastic apron, goggles ● respirator with consideration given to the chemicals and acids being used and appropriate filter packs for each ● well-ventilated room, plus ventilation hood under which to etch plates and exhaust fan with explosion-proof motor if flammable chemicals (resist developer, thinners) are to be used ● rosin powder for aquatinting ● asphaltum and Universal Etching Ground ● several small brushes ● paint thinner, lacquer thinner ● glass dish for resist developer solution ● several plastic darkroom trays for etching the plates (different sizes) ● red or yellow insect light (low wattage) ● sheet of plate glass to hold transparency in contact with resist-covered metal, a vacuum frame, or a contact-printing frame ● a handful of feathers (check along the shore of your local duck pond) ● Pyrex measuring cups in liter and half-liter (or quart and pint) sizes ● lightproof box for drying plates ● a hot plate ● small oven for baking plates ● butane torch ● paper towels ● an agitating device to assist in etching plates.

TRANSFER OF THE PHOTOETCHED IMAGE
FROM THE METAL PLATE

The embossed surface of the plate lends itself to image transfer.

Dennis Bookstaber, "Self Portrait Pin." Image etched into chrome plated copper, then fabricated with nickel, silver and brass. 1½" x 2½".

When the plate is sandwiched with a dampened pliable substance in an etching press, the plate will leave an impression whether or not it is inked. Sometimes extreme pressure isn't even necessary. Let us examine several of the possibilities available.

1. Inked Prints. Inking is relatively easy. Any photoetched plate can be inked, using a mixture of linseed oil and lampblack or other coloring agent. Commercially prepared etching inks can also be purchased.

Heat the plate slightly before inking. Apply the ink by rubbing it into the etched lines and/or aquatinted areas of the plate. Use cheesecloth, rubbing in a circular motion. Slowly the ink is wiped away from the flat, open areas. Do about four wipings, using fresh pieces of cloth each time. Whatever ink is left on the plate's surface will transfer to the print; so it is very important that the relief (highlight) areas be cleaned well. Sometimes the heel of your hand is useful for removing ink. If very small areas are holding an excessive amount of ink, removal can be accomplished with cotton swabs.

The image on the plate is now charged with ink and can be transferred to a variety of surfaces. The most common is 100% rag paper, such as Copperplate or Rives (BKF). First make the paper pliable by soaking it in water. Next, sandwich it between blotting papers and use a rolling pin to help remove excess moisture. Place the paper in contact with the photoetched plate, submitting both to tremendous pressure in an etching press in order to effect transfer of the image. These are the basic techniques employed by printmakers.

This process works with any impressionable surface. For example, one can transfer an image to a piece of leather or fabric, making soft sculpture out of the material. It is also possible to do multicolor inking, using different rollers and inks of various viscosities.

Inking and Printing a Photoetched Plate (left to right from top). 1. Ink the plate. 2. Handwipe the plate after 4 wipings of cheesecloth. 3. Roll on a second color. 4. Blot the soaked paper prior to printing. 5. The inked photoetching is transferred to the paper under tremendous pressure. See color section #1 for a full reproduction of printed image.

Heavier inks and softer rollers are used for more deeply etched recesses. Harder rollers and inks of less viscosity are used for the upper layers of the plate. In addition, it is possible to re-ink the plate with a different color and re-press the paper. Finally, different plates can be combined for montage image transfer.

2. Rubbings. Brass-rubbing effects can be created by placing a piece of paper over the photoetched plate and rubbing it with a pencil or crayon. Another option is to lay a moistened piece of rice paper over a deeply etched plate. The image is transferred to this delicate paper by rolling a brayer over the back of the paper. The image is given color by first loading the brayer with woodblock pigment prior to rolling it. In effect, the image is being created on the back side of the rice paper.

3. Ceramics. The etched plate can be used to transfer an image to a clay surface of leatherlike consistency. The impression of the image is then permanently baked into the final ceramic piece.

4. Plaster. Plaster prints can be made from a plate and displayed. Ink the photoetched plate and place it face up in a shallow box that will just hold it. Next, fill the box with plaster of Paris and allow it to dry. The image will be transferred into the plaster.

5. For Enameling. The plate can be employed to transfer etched images to a pre-enameled surface. Apply a thin coating of squeegee oil (a vehicle for silkscreening pigments) to the photoetched plate. Then lay a piece of waxed paper over the plate and rub gently with a burnishing tool to pick up the oil corresponding to the image. Next lay the paper, oiled side down, on the pre-enameled surface and transfer the oil image with gentle rubbing. Be careful not to smear the image. Now, dust powdered enamel over the image. It will adhere only where there is a coating of oil. Blow the surplus powder off. Once the oil is dry, the piece can be fired.

6. Embossing. Some experimentation has been done with pressing the photoetched plate into molten glass to transfer the image. The hardened glass will be embossed with the photographic image.

An uninked deeply etched plate can also create a beautiful embossed image on dampened 100% rag paper via an etching press (review 1., leaving out the inking step).

THE ETCHED IMAGE, AN OBJECT OF DELIGHT

Many different approaches to image presentation employ the photoetched plate itself as the center of focus. Biting completely through a plate and then mounting it was mentioned earlier as one alternative. Another is to enamel the plate. Not only would this be visually pleasing, but it could serve another purpose, as well: Any print edition previously made from the plate would thus be limited, preserving the scarcity value of the prints.

Images can also be presented in the form of photomurals. For those artists interested in enameling, metal sculpture, or stained glass, photomural fabrication offers a new realm of self-expression. Rather large murals can be manufactured from smaller sections or panels, utilizing the photoenameling or photoetching techniques described in this book. For example, a large piece of metal or glass could be cut up into manageable pieces to be photoenameled

Naomi Savage, "Silhouette." A photoengraving: deeply etched copper, silverplated, oxidized and lacquered.

Naomi Savage, "Portrait."
Production process the same as
"Silhouette" at left.

photoetched, electroplated, or whatever, and later reassembled. The effects will be most pleasing if the cutting lines have been incorporated into the more graphic lines of the image.

"Found" objects of metal, such as door knockers, light-switch covers, and metal buttons, could also easily be decorated with photoetched imagery. The possibilities, in fact, are endless: life-sized photoetched images on cars (mirroring one's personality); dinner bells, each with the portrait of a family member, each with its own sound; and so on.

DECORATING THE PHOTOETCHED PLATE

Prior to mounting the plate, consideration should be given to the coloring and the preservation of the piece. Here again there are several choices. Three areas will be discussed: The use of inks, chemicals, and fire to color and oxidize the surface of the metal; photoenameling alternatives; and electroprocessing alternatives, which will have a major section to themselves.

Martha Foster Banyas, "Kitty Box." Photosilkscreened underglaze "D" enamel on copper, kiln-fired. Ceramic butterfly decals and additional enameling was added in subsequent firings.

A. Inks, Chemicals, and Fire.

1. The plate can be inked in a variety of colors, using basic printmaking or painting techniques. The ink itself can be any one of several oil-based colors. Once the plate has been inked and dried, it can be coated with a clear lacquer to protect the metal from oxidation.

2. Certain chemical agents and oxidizers can be employed to oxidize and color a plate either completely or selectively. **a.** For example, silver and copper will both turn black when treated with liver of sulfur (sulfurated potash). Make a working solution by adding about five grams of powdered liver of sulfur to 960ml of hot water. Apply the solution with a brush for localized application, or cover the piece completely by soaking it. The metal will darken rapidly. Next, place the piece in cold water for one minute. Do not touch the surface while it is still wet. Once the piece has dried, buff out the highlight areas with jeweler's rouge. Finish up by applying a protective coat, spraying with a glossy lacquer.

b. Rapid Selenium Toner and Poly Toner (both from Eastman Kodak) will also oxidize and blacken metals such as copper and zinc. The metal tends to oxidize initially in multicolors, followed by gray tones, and then go deep black, depending upon the dilution of the toner.

Several other colors are possible. **c.** For a green color on copper, try adding 6 grams of copper nitrate to 720ml of water. Heat the solution and brush it on, applying several coats. **d.** To obtain a brown color on copper, dissolve 10 grams of copper sulfate in 20ml of water. Heat this mixture and apply it when it is hot. Clean the piece with pumice powder and water, then reapply the solution in order to obtain a uniform coating. Keep cleaning and reapplying the solution until the desired tone is reached. **e.** Lastly, ferric chloride itself may be employed as a colorant, since it blackens zinc, copper, brass, aluminum, etc., while it etches the metal. After etching the highlights can be buffed out, with no further treatment necessary. Again a glossy lacquer spray will protect the surface of the plate from scratches and from further oxidation.

3. Fire can be employed creatively to produce a rainbow of colors on metal by burning and oxidizing the surface. For example, copper, when heated with a torch, turns bright red and then greenish-blue and blackish. With some control one can obtain a nice mixture of colors on the surface of the metal plate: red, orange, yellow, green-blue. Use a pointed flame in order to control the areas of heat application. When the piece looks right, immerse it immediately in water. The resulting effect works nicely with some images. Finally, use lacquer to preserve the coloration. A variation on the above technique would be to immerse the hot plate in motor oil instead of in water.

B. Photoenameling Alternatives.

There are several methods for incorporating images of enamel or pigmented glass into metalwork.

1. Champlevé Inlay Technique. On a piece of copper or silver that has been photoetched reasonably deep (0.8mm to 1.6mm—1/32 in. to 1/6 in.) it is possible to pack the recesses of the image with enamel. The piece is then fired, using conventional enameling technique (described below). Not all metals can be used. Zinc, for example, has a melting point below the fusing point of the enamel itself. Generally, use of a heavier gauge of metal is recommended, as thinner gauges tend to soften too much during firing.

Either transparent or opaque colors may be employed. All enamels must be washed in distilled water prior to use. In one method, the pigmented glass is mixed while still wet with some enameling gum and wet-packed into the recesses of the plate with a tiny spoonlike spatula (dentists' tools work well here). The enamel should be heaped slightly above the etched depressions because the firing tends to make it settle. Once the enamel has dried, the plate can be fired. Another method involves covering the entire piece with a fired-on enamel top, then using a set of silicon carbide stones, several grades of sandpaper, and finally a fine polish of cerium oxide (optical grade) to smooth down the enamel surface, re-exposing the metal in the nonetched areas while leaving enamel embedded in the etched depressions. The exposed metal is then buffed with jeweler's rouge. Images with wide expanses of shadow area will not work well with this technique. In either of these methods, be certain that enough enamel is applied or else the image will go black.

Selective application of an oxidizer to a photoetched piece of zinc.

Prior to placing any enamel on the metal, prepare the surface to receive it. Remove oxidation by gently heating the metal in the kiln to a dull red glow. Next, place the piece in a pickling solution of Sparex No. 2 to clean the surface chemically (review precleaning of metal surfaces in photoresist section). Make certain that all photoresist and ferric chloride residues are removed from the image area. Clean the plate in running water and dry it with a paper towel. Use a glass fiber brush to remove any scale caused by firing.

In most instances the piece should be counterenameled, or reinforced, to discourage warping and cracking of the enamel surface. This should be done now. First, paint a coat of Scalex over the etched side of the plate to protect this surface during firing. Then apply a coating of gum or Klyr-Fire on the side to be counterenameled, followed by a coating of enamel from an 80-mesh sifter. Dry the piece on top of the hot enameling kiln. Next, place it on a trivet and insert both into the kiln. Fire the piece, using the temperature chart shown here. When the piece is removed from the kiln, it is set aside to cool. If it has warped, however, replace it in the kiln and refire it with the concave side up. As it is reheated it will begin to slump back to a flat position. When it does, remove it from the kiln and quickly place it on a cold, flat, steel surface. Quickly, gently yet firmly, place a heavy object—such as an old household iron—on top, pressing downward until the enamel cools. With care the piece can be saved.

Enamel can now be applied to the photoetched surface, dried, and fired. It is very important that the work area be free of dirt, to avoid getting fire scale or dust on the piece. These will show up as undesirable black specks.

Enamel colors always look different after firing. It is therefore a good idea to run tests of colors on clear and on white backgrounds before doing the actual piece. Thompson's No. 1005, Medium Fusing Transparent Flux for Copper, and Thompson's No. 621-A, White Undercoat for Transparents, can be used for this purpose. When running tests of enamel paints, also try ceramic overglaze pigments used for china painting, as they, too, can be fired on enameled surfaces.

For best results fire enamels by temperature, using a pyrometer. Most enamels, unless otherwise specified by the manufacturer, are medium-fusing; that is, they melt between 677° C and 843° C (1250°-1550° F). Hard enamels are usually fired first, followed by subsequent firings in order of enamel softness. It is possible, however, to fire medium-fusing enamels over soft enamels, if the two layers are separated by a transparent flux coat. In firing, the surface of any piece will first turn black and then become granular. As soon as the enamel turns molten red hot and glossy-looking remove the piece from the enameling kiln. This is the point just beyond the orange-peel effect (slightly rippled surface) where the enamel smooths out.

Kiln Temperature Chart

hard enamels	*843° C (1550° F)*
medium enamels	*816° C (1500° F)*
soft enamels	*788° C (1450° F)*
commerical decals,	*677° to (1250° to*
enamel paints,	*740° C 1300° F)*
china paint,	
lusters and	
overglaze colors.	

Generally, the surface of the piece will first turn black and then become granular. As soon as the enamel turns molten red hot and glossy-looking, remove the piece from the enameling kiln. This is the point just beyond the orange-peel effect (slightly rippled surface) where the enamel smooths out.

After the piece is removed from the kiln it is cooled slowly on an asbestos sheet. (Do not dump the piece in water, because the enamel will certainly crack.)

The piece may now be cleaned. Remove any fire scale with household cleanser (Ajax) and water. Next, smooth the enameled surface by rubbing with a fine emery stone under water. Set the piece on a scrap of old leather to avoid damaging it while you work. This is the point at which to repack and refire the piece, if it is necessary to build up the enamel deposit.

Once the piece is ready for its final polishing, follow this procedure: Use a fine silicon carbide stone, then a Scotch stone (after stoning always use a glass fiber brush to remove residual particles); next, to get the surface really smooth, use wet or dry silicon carbide sandpapers in the following order of coarseness: No. 180-grit, then 280-, 320-, 400-, and 600-grit. The coarsest is 180-grit, and 600-grit is the finest.

Once the piece has been fired and cleaned, exposed areas of metal can be buffed and polished, selectively oxidized, and/or electroplated. The plating solution will only adhere to those areas free of pigmented glass. It is also possible to work the piece further by photoetching the enameled surface itself, using hydrofluoric acid (extremely hazardous, CAUTION is needed) as the etchant. Secondary images created in this fashion would have a matte quality, in contrast to the gloss characteristic of enameled surfaces. Wax is a useful resist for touch-up work whenever hydrofluoric acid is employed.

2. Photosilkscreening Pigmented Enamel. Several options exist for using pigmented enamel and a silkscreen to add photo imagery to metal. The Thompson Company markets a product called Underglaze D. This is a highly pigmented black oil-based enamel that can be photosilkscreened directly onto bare metal. Before applying it, let the paint thicken by leaving the lid ajar, so that some of the solvents can evaporate. When screening this compound it is best to use about a 157 to 200-mesh multi/filament polyester, so that particles can pass through the screen easily. A finer mesh can be used, but at the sacrifice of detail; this is the typical dilemma of any photosilkscreen operation. (For more on Photosilkscreening technique, see Chapter 9.)

image. Once you are satisfied with the quality of the image, dry the surface and fire the piece. Unlike other enamels, Underglaze D does not become shiny on firing. In addition, after an initial firing the image and the metal can go through a gentle stretching, to dome or otherwise alter the surface. The manipulated piece is then sifted over with a coat of clear flux and fired again.

At this point, designs in transparent enameling paints can be added over the encased image, using water or Klyr-Fire to hold the enamel to the surface of the piece. In addition, photodecals made with enamels can be used to embellish the piece further; however, it is very important that all decals be applied just as they slide off the sheet. They must not be flipped over prior to application.

There are several other possibilities for photosilkscreening enamels. For example, steel tiles that are commercially coated with white enamel top or copper sheeting can be cut and bent to any desired shape and given a hard-firing, a white enamel base-coat. (Note that a warm white is preferable to a cold white basecoat.) Any such surface will accept a photodecal or a photosilkscreened image formed in pigmented enamels. Mix the enamels with squeegee oil to a screening consistency. The oil will fire out with no visible residue.

3. Dusting-On Enamels. Derivations of the dusting-on process (described in Chapter 7) were used in the past to create glass images on porcelain surfaces. Today experimentation with them appears worth while. With this process, under subdued light a tacky emulsion consisting of a sensitzed gum-and-sugar solution is coated onto a pre-enameled white surface. The emulsion may be thinned with water, if necessary. A positive transparency is laid over the emulsion, which is then exposed to UV light. The light hardens the exposed surface while the unexposed areas remain tacky. The image is "developed" with enamel paint ground to a fine powder, which adheres only to the tacky areas. The image is then fired for permanency. Several reapplications of enamel, with firing between each layer, may be required in order to get a strong image.

4. Pigmented Direct Photoemulsion. Some experimentation has been done using a directly applied photosensitized emulsion loaded with china paint or pigmented enamel. Promising results have been obtained. Images made in this manner by-pass screening operations and permit images to be placed more readily on large or curved surfaces. Major findings indicate that lighter coatings of emulsion give better results with less chance of flaking or pitting. For instance, when a very thin coating is used, 65-line halftone images can be fired successfully. To make an emulsion use:

• 1 tsp PVC-PVA (polyvinyl-chloride/polyvinyl-alcohol) or a direct photosilkscreen emulsion (the base)
• ½ tsp Phenoseal (a commercially prepared vinyl compound) and
• ½ tsp white glue (Elmer's)

Kent E. Wade, "Deeta's Backside." A photoengraving: a zinc plate was copperplated, then photoetched, handworked, oxidized and lacquered. 4" x 5".

• ¼ tsp ammonium dichromate sensitizer, stock solution (1 part ammonium dichromate to 5 parts water)
• ½ tsp enamel powder (color of your choice).
Combine the base and the additives, then mix in the enamel. Working under subdued light, add the sensitizer when you are ready to coat the piece. Apply with a small brush in two very, very thin layers. Dry between applications with a hair dryer. Always use metal with a fired-on white undercoat (for optimum image contrast) or a clear flux base. Roughen the surface slightly by stoning the undercoat in order to increase adhesion of the emulsion.

Place a negative transparency in contact with the light-sensitive emulsion, expose to ultraviolet rays, develop in water, dry the surface, and kiln-fire the piece. Use a lower firing temperature than required to fuse the enamel, in order to burn out the emulsion first. Then increase the temperature to the range specified for the enamel. The emulsion will have burned away, leaving the enamel fused permanently to the metal in the shape of the image. *BEWARE* of poisonous fumes from the polyvinyl chloride in any of the base substances used.

Other colloidal vehicles, such as gum arabic and egg albumin, need to be evaluated. They should be able to hold reasonably large quantities of enamel and yet burn away without leaving ashes. (See the chapters on photoceramics, nonsilver processes, and halftone transparencies.)

CHECKLIST OF ENAMELING EQUIPMENT AND MATERIALS
• metal (18-gauge or thicker) • electric enameling kiln •

Direct Photoenameling Process (left to right from top). 1. Coating a copper enamel surface with a light sensitive, enamel/based emulsion. 2. Exposing the sensitized plate. 3. Develop the piece. 4. Dry the image prior to being kiln-fired. 5. Removing the fired photoenameled piece from the kiln.

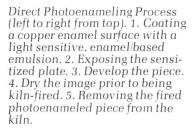

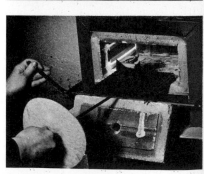

pyrometer • rubber or insulated floor mat • metal cleaner (Sparex No. 2) • protective coating (Scalex) • binder (Klyr-Fire) • basecoat, medium-fusing white • enamel, enameling paint (bottles for each) or ceramic overglaze (china paint) • transparent flux (optional: special underglaze for silkscreened images) • direct and indirect photoemulsions (optional) • silkscreen medium (squeegee oil) • firing tools: trivets, asbestos mitten, firing fork • 80-mesh sifters for enamel application • inlay tools • glass-fiber brush • household cleanser • graded sandpapers, silicone carbide stone, emery, Scotch stone • jeweler's rouge, buffer, buffer wheel • leather scrap (for working surface).

ELECTROPROCESSING ALTERNATIVES

Of all the tools available for decorating metal, few are as fascinating as the electroprocesses. Photographic images can be created electrically in multiple layers of dissimilar metals. When all is done tastefully, other decorative techniques can also play a role in image presentation—enameling, selective oxidation, and so on.

A. What Are The Electroprocesses?

They are simply methods for transferring metal in a highly conductive bath (electrolyte) through which low-voltage direct current flows between a positive and a negative pole (electrodes) at opposite ends of the bath. The current can originate from sources such as small dry-cell batteries, a car battery, or from standard household current (explained in part E. below). We shall look at three major processes.

Kent E. Wade, "Untitled." Kiln-fired, photo-enameled image using the pigmented, direct emulsion process. Copper plate was first enameled with a white undercoat base. Ceramic overglaze was employed. 2" square.

Electroplating adds metal to the surface of an object by electrolytic means when that object is connected to the *negative* terminal. This involves passing an electric current through a special bath (see part D. below). If a resist is employed on the surface, then plating, or electrodeposit of metal, will take place only in the unprotected areas. Light-sensitive resins and other substances, such as asphaltum, are suitable plating resists. Electroplating can be accomplished by immersing the object in a bath or done selectively with the aid of a brush-plating unit (see part F. below).

Electroetching removes metal by reversing the position of the object in relation to the electrical flow required for electroplating. Here, the object is secured to the *positive* terminal. A different bath than that required for plating is recommended. The function of a resist in this technique is to prevent removal of metal. Photoetching with electricity, therefore, provides an alternative to photoetching with acid. Electroetching is also a useful technique for brightening metallic surfaces by removing heavy oxide deposits.

Electroforming is a method whereby almost any *nonmetallic* object can be plated; that is, a metal object is formed around a nonmetal core. Thus highly individual shapes can be created and used as substrata for photoetched or electroplated images (see below, B. and C.).

B. Metal Color Contrast

Here are a few possibilities for creating color contrast with metal:

1. A copper base can be plated with nickel, silver, or gold. Many metals will not adhere to each other, although most metals adhere

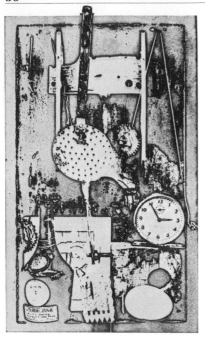

Steve Amen, "Still Life." A photo intaglio print. Hand wiped in brown and surface rolled in light blue. 12" x 16".

well to copper. In addition, copper is inexpensive in comparison with other metals; this is why it is usually chosen as the substratum or as an underplate in the multilayering of metals.

2. A zinc base can be plated with copper. Try a copper cyanide bath to coat the zinc and then photoetch or electroetch through the copper. *WARNING:* Cyanide forms a poisonous, odorless, yet deadly gas if it comes into contact with any acid. Make sure there is good ventilation when using the two chemicals in the same area.

3. There are also black-gloss and black-antique plating baths available for electroplating metal surfaces. These "metals" bond with the substratum to create an oxidized effect, unlike the liver of sulfur treatment, which simply discolors the metal's surface (review decorating the photoetched plate). The baths are expensive, and a special anode is required.

4. There are many other electroplating solutions available for adding color to metal. For example, a lead plating bath can be used to produce a royal red on copper. (*WARNING:* Lead fumes are very dangerous, as is the potential for lead poisoning.) To make the electrolyte, dissolve 180 grams of caustic soda into a liter of deionized water (Solution A). Next, dissolve 30 grams of lead oxide and 7 grams of gum arabic into 3 liters of water (Solution B). Slowly mix A and B together. Warm this combined solution on a hot plate to approximately 54° C (130°F) and use with 2½ volts. The deposit of lead can be monitored by periodic removal and inspection. Once the copper appears to be coated satisfactorily, heat the piece in an oven to a cherry red, then let it cool. Next, buff off the residual black coating to expose the royal copper color.

5. A heavily worked photographic fixer can also be adapted for use as a silver-plating bath for copper.

6. In order to achieve a variety of color contrasts, the order of plating steps may be varied:
- a. electroplate before applying the photoresist;
- b. apply the photoresist and then electroplate;
- c. photoetch the plate and then electroplate the recesses;
- d. plate and combine steps selectively.

For example, a sheet of copper could be plated with silver and electroetched (by electrolysis) or photoetched (with acid) to create an image of a yacht and a sunset in copper, with a silver background. Next the yacht and the background could be coated with a metal resist, such as asphaltum, leaving the sun unprotected, to be plated in gold. A frame could be designed out of cardboard or wax and electroformed in yet another metal and placed around the image.

7. Electroplating, electroetching, and electroforming techniques could be combined in one piece. For example, a wooden object could be plated with copper by electroforming technique, then an image could be plated on the copper surface with a dissimilar metal, such as nickel. Still another image could be added by electrically etching other areas.

C. Electroforming Preparation

When wood is to be electroformed, its pores must first be sealed with a varnish or wax. A highly conductive wire is then attached for use as an electrical contact point. (It will be removed after the piece has been plated.) The piece is cleaned with a degreasing compound and hot water, followed by application of a metallic medium, such as a silver-based conductive paint. A homemade medium can be made by combining 12ml of lacquer thinner with 14ml of clear lacquer, then adding approximately 28 grams of a fine conductive powder, such as a copper dust. This mixture is painted onto the surface of the wood. If the initial coat has a glossy appearance, the solution may be diluted slightly with more lacquer thinner before a final second coat is applied. After this has dried, there should be a frostlike film covering the surface. The piece can now be formed by plating it in a copper electrolyte (see part D.). An image in nickel could be added to the new copper surface by using a photoresist and electroplating. Alternatively, the copper surface could be nickel-plated, then photoetched to re-expose the copper in the form of an image. (Electrical setup and cautions on using it are under heading E. below. See G. for methods.)

D. The Electrolyte

An electrolyte is a conductive chemical bath prepared for ion transfer of a specific metal. The bath is selected on the basis of the metal that is to be transferred and whether you are plating with it or electroetching a surface coated with it. For example, to plate copper with silver will require a silver-plating solution, whereas a copper substratum already plated with silver will require a different solution to electroetch the silver layer.

Kent E. Wade, "Goat Lady."
A photoengraving: Photoetched copperplate, image was then nickel plated and lacquered. 4" x 5".

Each bath works best when the electric current is maintained within a certain range. Current is measured in amperage and is a function of voltage, bath temperature, and bath chemistry. Control of these variables is important for precise work; otherwise plating times and the quality of the plate will be affected. Baths are generally designed for use at a slow plating rate in order to achieve an even deposit. Electrolytes used at rates of current above or below what is recommended will still plate metal, but less smoothly. For example, prolonged plating at excessive current tends to produce an uneven deposit with a granular effect. Variations like this should be mastered for their aesthetic usefulness.

Commercially prepared electrolytic solutions are available ready for use from jewelry supply houses. They are presently packaged in one-quart plastic bottles, priced according to the basic metal component. For example, nickel and copper baths are relatively inexpensive, while silver-based and gold-based baths are more costly.

Most premixed electrolytes will plate an object within one-to-five minutes. Some of these solutions, such as H. R. Superior's Immersion Gold, will even plate without external power. With this bath, the object can be plated within 40 seconds merely by immersing it, attached to a special activating wire, into the heated solution.

Plating solutions can also be made at home. For example, a copper-plating bath can be made by dissolving 190 grams of copper sulfate into one liter of deionized water. Slowly add 50ml of sulfuric acid. Heat the solution to approximately 30° C (85° F), using it at two volts, ½ ampere. (Other solutions were mentioned earlier under B., metal color contrast.)

Note: A longer bath life is attained when deionized water is used for mixing an electrolyte, although distilled water may be substituted.

Cyanide-based electrolytes should be avoided when electroplating images. Electrical action in these baths will liberate hydrogen gas at the cathode, and this causes the photoresist to lift off the metal.

For the same reason, alkaline-based electrocleaning solutions should be avoided. In order to degrease and deoxidize the nonresist areas before electroprocessing, immerse the object in a 10% solution, by volume, of hydrochloric acid or a 10% solution of ammonium persulfate for about 30 seconds. Next rinse the object in distilled water. If water beads up on the surface of the object, reclean it in acid. The object is clean when water will cascade across the surface in an unbroken, uninterrupted flow. Most failures in electroplating can be retraced to improper cleaning; so proceed slowly.

E. Electricity for the Electroprocesses

As previously stated, in electroprocessing a current flows through a conductive bath between two electrodes, the anode (positive) and the cathode (negative). The piece to be electro*plated* is attached to the *cathode*; for electro*etching*, it becomes the *anode*, instead. As electricity flows through the chemical bath (electrolyte), it stimulates molecules of one metal element to combine with those of another, forming a permanent bond yet leaving the color of each metal distinct. Resists will prevent the transfer of ions, however, and this is what makes possible selective plating—in this case, the formation of an image.

A Word of Caution: *When you are working with acid, water, and electricity, be sure to wear rubber gloves, stand on a nonconductive surface such as a rubber mat, and make certain everything is well grounded. A plastic apron and goggles are also advised, in case solutions are accidentally splashed or spilled. Be sure clips holding anode and cathode wires to rectifier do not touch the electrolyte solution. Use insulated wire.*

Direct current is necessary for electroprocessing. It can originate from D.C. batteries, or from any 115-volt alternating-current outlet with a D.C. converting rectifier. The rectifier also provides variable control over power output, which is required for plating with different solutions. Voltage requirements are low, from two to ten volts, regardless of object size or bath size. The significant measurement is amperage, rather than voltage, because the surface area that can be plated and the size of the plating bath are directly related to the unit's output of amperes. Amperage is a measurement of the volume of the electrical current, as opposed to voltage, which is simply the electric pressure needed to maintain the flow of current. Rectifiers can be purchased from jewelry or electrical supply houses. Prices will vary with the amperage output of the rectifier.

A trickle charger that produces about one ampere at six volts can be used to plate small surface areas, up to 10x13cm (4x5 in.). For larger areas, one could use a rectifier that converts household power to direct current, has variable voltage output up to ten volts, and will produce at least ten amps. This is plenty of amperage to plate or etch a 20x25cm (8x10 in.) surface area with most plating solutions. For larger areas, a 25-amp rectifier or the services of an industrial plating

Electroplating Equipment.

firm is recommended. In the latter case, offer to deliver the metal ready for plating in order to defray some of the labor costs. This would include preparation of the image in photoresist form and attachment of wires for electrical hookup.

A ten-amp variable-voltage rectifier can also be made at home. A well-grounded ten-amp battery charger with a built-in circuit breaker is plugged into a rheostat (voltage-control device); the flow of electrical current is monitored with a voltmeter and an ammeter (measure of amperes) tied into the electrical circuit. A dimmer switch can be used as an inexpensive rheostat, but be certain that it can handle the amperage or it will burn out. When building a system, it is best to seek the advice of someone knowledgeable about electricity. There's no point in getting all charged up over nothing!

Electric current flows from the positive to the negative terminal; therefore it is the anode that introduces the positive charge into the electrolyte. The anode should, if possible, have a surface area identical to that of the piece being plated. This helps to make the plating more consistent in thickness. For general work, a larger anode can be used to create a greater flow of current and thus a faster deposit.

The anode is normally of the same metal as that to be plated onto the piece. The exception is when a stainless steel anode is used. For example, when plating copper with nickel, a nickel anode is recommended as it dissolves into the electrolyte (a nickel solution, in this case), replenishing the nickel being drawn out of solution and deposited on the copper. If a stainless steel anode is substituted for the nickel anode, nickel will be taken out of solution without being replaced. Eventually, the bath will become ineffective.

F. Brush Plating

Brush plating permits selective plating of small areas without the inconvenience of setting up a larger electrolytic bath. Hand-held battery-operated units can be obtained from jewelry supply houses. Electrolytes are also available packaged in small two-ounce bottles. They are specifically formulated to be high in metal salts.

If you have access to a rectifier, a brush plater can be made easily at home. Wrap a copper wire around the bristles of a small brush. (Use a brush without a metal ferrule.) The handle of the brush should be made of wood or some other nonconductive material. The opposite end of the wire is attached to the positive terminal of the rectifier. The brush becomes the anode. The object to be plated is connected to the negative terminal of the rectifier by another copper wire. To use the brush plater, dip it in the electrolyte and "paint" the areas not protected by any resist coating. Repeat this process until the desired coverage is achieved. Do not touch any of the wires during plating.

A brush plater can also be made by running a copper wire through a glass tube and into a sponge wedged into one end. The tube acts as a reservoir, being filled with electrolyte and corked at the opposite end.

When building a brush plater, *make certain everything is properly grounded and well insulated.*

G. Basic Steps in Electroplating/Electroetching

1. Prepare metal object with photoresist stencil of image.

2. Select appropriate electrolyte and anode.

3. Heat the electrolyte as recommended (using hot plate and darkroom thermometer).

4. Solder one copper wire to the anode and another to the object; both wires should be long enough to extend out of the bath. should be long enough to extend out of the bath.

5. Degrease and deoxidize both the object and the anode and rinse in distilled water (review last paragraph under D.).

6. Pour the heated electrolyte into a container large enough to hold both the object and anode. (One-liter beakers can be used for small plates and objects; acid-proof plastic buckets or trays are fine for larger pieces, as plating can be done in either horizontal or vertical position.)

7. Place the object and anode into solution so that both are completely covered.

8. IMPORTANT: Before proceeding, check that the rectifier is *properly grounded* and *switched off.*

9. Check that the insulated wires are connected to the rectifier with alligator clips at opposite ends: one wire to the negative terminal, one to positive terminal. For **electroplating,** clip the object wire to the negative line to make the cathode and clip the anode to the positive line. For **electroetching,** reverse the above electrical connections, making the object the anode (review A., E.). *DO NOT LET alligator clips touch the solution.*

10. Set the rectifier for the recommended voltage and timer for the recommended plating time.

11. Turn on the power unit.

12. If agitation is required, stir bath chemistry with a glass or plastic stirring rod. Wear rubber gloves.

13. If the anode becomes coated with impurities and the plating process slows, turn off the unit and reclean the anode by rinsing it in distilled water. Return the anode to the bath and resume plating.

14. When the specified plating (or etching) time ends, turn off the rectifier, disconnect the object and remove it from the bath.

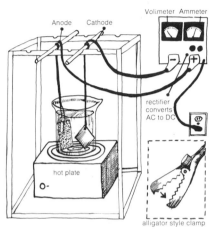

An Electroplating System.

15. Wash the object in running water, dry, and examine it.

16. If a heavier plating (or etching) is desired, reclean the object, replace it in the bath, recommence process.

17. When the process is complete, filter the electrolyte before storing it.

Note: The electrolyte and the anode can be stored for reuse; however, the solution should first be paper-filtered to reduce contamination from process-related impurities. It can then be stored in a plastic container. If the bath has become so contaminated that it has lost its plating ability it should be discarded, but not just anywhere. Check with local city agencies regarding disposal procedures in your area.

If desired, the object can be reworked using other techniques, such as enameling and photoetching. The wire attached to it can be removed at any time.

Dennis Bookstaber, "Untitled." Photofabricated, laminated belt buckle. Image photoetched on chrome plated copper, fabricated with layered brass to form buckle. 2½" dia.

ELECTROPROCESS EQUIPMENT CHECKLIST

• DC power source • a *well-grounded* rectifier • voltmeter, ammeter, and variable-voltage control • a selection of plating solutions and the proper anode for each • 18-gauge copper wire (insulated) and two strong alligator clamps (anode and cathode wires) • acid precleaning solution • good-quality darkroom thermometer • hot plate • Pyrex beakers and/or plastic containers large enough to hold object and anode • distilled or deionized water to replenish evaporated plating solution and for purification rinse step • glass or plastic stirring rod • rubber gloves, plastic apron, goggles • a suitable photoresist and developer • timer

PRESENTATION OF PHOTOGRAPHICALLY WORKED METAL SURFACES

There are numerous ways to present photoetched, enameled, and electroprocessed images. A few ideas are mentioned here.

• Metalwork can be mounted with epoxy glue directly onto wood or other surfaces.

• A wooden or metal picture frame can be used, with the photomanipulated plate replacing the glass in the frame.

• Contrasting the metal with wood and fabric can strengthen the presentation of the image. Secure a picture frame large enough to leave a substantial margin around the metal plate. Next cut a solid wood or masonite backing to fit in the frame. Cover it with black velvet or another fabric that complements the colors of the image and the plate. Attach two small bolts or strands of wire to the back of the plate with solder. Center the plate on the fabric-covered mounting board and bolt or wire it through the material and the board. The mount is then placed in the frame, where it is held snug with small nails. This system permits the frame to be reused for similar-sized plates, since each plate or mounting board can be removed and replaced easily.

• The metal plate evokes a pleasing visual response when raised off its display surface. A raised appearance can be created by using silver-colored mirror clamps at each corner of the photoworked plate. These permit the plate to stand about 13mm (½ in.) off its support.

• Frames can also be manufactured out of metal. For example, frames can be made in copper, applying silver solder at the joints. Next, the frame itself can be electroplated with nickel or silver. Bezel (a thin ribbon of silver) may also be employed effectively to create a metal enclosure for photoengravings or electrically etched work in jewelry applications.

More than likely, once the techniques are under control you will be integrating photoetched, enameled, or electroprocessed images into other works and fabricating objects of metal in which the image plays a subtle role. Thus your needs for framing will change, and experimentation will be necessary.

George Johanson, "Flesh Morning #1." An intaglio print. A zinc plate was made using photoetching, soft ground etching and aquatinting technique. The plate was inked via handwipe, roll-on and à la Poupèe methods. 22" x 16".

Photo-presentation on Glass

Photographic images presented on glass surfaces can be very effective. The translucent and reflective qualities of glass give the image a new dimension characteristic only of this medium. There are many different methods for affixing the image to glass, and each is unique. Five general areas will be discussed in this section: 1. etching images in glass with acid; 2. sandblasting images on glass surfaces; 3. photosilkscreening images on glass; 4. using special photosensitive plates, including photoemulsions for glass applications; and 5. presentation, or display, of the glass images. One area that will not be detailed is photodecals for glass; instead, see the section on photoceramic decal fabrication (Chapter 3) for a description of this technique and its basic principles.

TYPES OF GLASS FOR ETCHING, SANDBLASTING, AND SILKSCREENING

There are several types of glass available, each with its own special characteristics that can enhance image presentation:

1. Window and colored glass: Clear, single- or double-strength window glass or plate glass, which is even thicker, are the most commonly used. In addition, there is a variety of colored glass (stained, tinted, etc.) available. Experiment with tinted, smoky-colored glass. Matte-etched or sandblasted areas of an image on this glass will appear white in contrast with the darker background, improving image visibility.

2. Flashed glass: Flashed glass is particularly interesting. It consists of a thin layer of colored glass baked onto one side of a sheet of clear glass. It is also available with two colored layers superimposed one upon the other. With flashed glass, one can etch through the colored layer and expose the clear glass, or selectively sandblast the colored layer away, creating a white translucent surface. Matte etchants can also be used; however, they will only frost flashed glass, and will not etch through the colored layer.

Richard Posner "His Masters Voice." Positive transparencies sandwiched between clear glass, wrapped at edges with copper foil, soldered and incorporated into the stain glass picture window.
(Pat Goudvis; Photo credit)

Halftones, as well as line images, can be reproduced on flashed glass. Halftone images actually appear to have shades of tone in the colored layer, with the extreme highlights revealing themselves as completely clear or frosted glass, depending upon the process employed. If you attempt a halftone image, use a negative transparency. For simple presentation, remove complex backgrounds around the subject, cutting the transparency with a sharp mat knife or one-sided razor. The image will appear, after etching, to have a colored background.

Unfortunately flashed glass is expensive; therefore, consider making your own. Colored glass frit can be screened onto clear glass and kiln-fired to specifications (see section on fired-on glass images).

Sometimes it is hard to distinguish between the coated and uncoated sides of flashed glass. Just remember that the usual sound associated with cutting glass will be absent when you cut the flashed side of the glass. Generally, for cutting glass use a Fletcher's No. 02 cutter. Smear a drop of oil on the cutting line to insure a cleaner cut.

3. Frosted glass: If a piece of glass is frosted slightly on one side with sandblasting apparatus or a matte etchant, acid will bite through the frosted ground and produce a clear, nearly transparent image amid the frost. This effect can enhance the image, as the glass is still translucent but not quite transparent. Some images will definitely show up better on translucent surfaces, as the viewer will not be distracted by physical objects that might be seen through more transparent glass. It is worth noting also that a photosilkscreened ink, or a photoresist, will adhere excellently on the roughened surface of frosted glass.

4. Enameled and mirrored surfaces: An enamel is actually a glassy substance that can be fused to metal. Once an enamel coating has been fired on, it can be photoetched like any other glass, using hydrofluoric acid as the etchant. Finally, a mirrored surface may be photoetched or a glass surface selectively mirrored for added visual impact.

ETCHING PHOTO IMAGES IN GLASS WITH ACID

The intricate detail achieved with the use of acid on glass is unexcelled; however, this is by far the most dangerous of the techniques we shall discuss. Hydrofluoric acid and formulas containing it are the only compounds, or etches, that will dissolve glass. This acid is potentially very dangerous to human beings in that its hazardous vapors will attack the respiratory system, the acid itself causing severe burns. Most acids can cause burns, but they do so immediately; hydrofluoric acid may not burn or cause pain upon contact, and thus one is unaware of the need to neutralize its action until living tissue has been "eaten." This may sound gruesome, but it is meant as a caution to treat this acid with the respect it deserves. It is very important to use only plastic trays, to wear rubber gloves, eye protectors, and a plastic apron, and to work either out of doors or in a place that has superior ventilation. Keep neutralizing agents close by: Baking soda will work for skin contact, and Phillips' Milk of Magnesia is said to be gentle enough to put in the eyes. Again, be careful with this acid!

June Marsh "Untitled." Two Kodalith transparencies of original drawings were individually sandwiched between sheets of clear glass. The images were then set one inch apart in a 2" x 4" wood frame to create movement in the piece when viewed from various positions. The backing sheet of glass for the rear image is sandblasted, diffusing the light source.

Hydrofluoric acid, full strength or in a diluted state, etches glass clear. Hydrofluoric compounds, such as Screen Etch, give a white matte effect. Each formula has its use and produces different visual qualities.

A. The Use of Hydrofluoric Acid (Clear Etch)

In order to etch glass, an acid-resistant image must first be created. There are three basic methods.

1. Photosilkscreen technique: This technique can lay down an image in asphaltum, wax, or a plastic compound, any of which will resist the acid's biting action. Once the resist has been applied, the back and edges of the glass should also be protected. Use contact paper or a thick coating of asphaltum on these vulnerable surfaces.

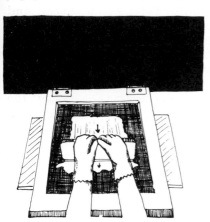  

(Left to right). **1**. Photosilkscreening a resist onto glass. **2**. Coating the back of a piece of glass with asphaltum prior to etching. **3**. Etching technique for glass (string travels underneath glass to set it in the acid.)

2. Photoresist technique: This can be used to create an image in stencil form on any glass surface. Halftone and line images can be etched successfully into glass using G.C. Photo Resist Sensitizer. Other resists can also be used and may even adhere better: try Kodak KTFR or Dynachem 5000 Resist.

Once an image in the resist medium has been reproduced on the surface, the glass can be immersed in a bath of hydrofluoric acid (go slowly to avoid splashing). On extra-large pieces where it is difficult to obtain a container big enough to hold t e glass, it may be more feasible to build up a wax-resist lip around the perimeter of the piece first, and then pour the acid bath slowly over the surface.

Begin by experimenting with a solution consisting of one part hydrofluoric acid to two parts water, always pouring the acid into the water (never pour water into acid). It is unnecessary to agitate the piece in the bath. Indeed, it is wise not to because without adhesion aids, the photoresist tends to lift off. Etching is accomplished easily within 5-to-20 minutes. Develop, using the periodic removal-rinse methods outlined for etching metal (see Chapter 1). Observe the finer lines and the halftone highlights carefully. With prolonged etching, these image areas disappear first. Remember that hydrofluoric acid should only be used in plastic containers, as it will attack metal as well as glass. The process outlined above is the basic method. It is not a difficult procedure; it is just a dangerous one and caution should be exercised.

The precleaning and post-process-baking stages of the photoresist technique are particularly important when etching glass (review photoetching of metal). The glass surface must be free of dirt and oil if the resist is to hold properly and the acid to etch evenly. In addition, images formed in photoresist must be baked on to avoid premature liftoff and undercutting by the acid. When heating the glass, be particularly careful of thermoshock: Heat and cool it, both, slowly and evenly.

Adhesion of the photoresist can also be improved by using special intermediate layers or bonding agents. There are several alternatives.

a. For *deep-etching* of glass, an epoxy catalyzed paint is advised as an intermediate layer. The photoresist clings far better to the intermediary than to the surface of the glass; thus, undercutting of the resist and the possibility of liftoff are lessened. The photoresist is coated over the epoxy, exposed and developed as usual, and given a short postbake. The epoxy paint must then be removed from the resist-free areas with a solvent that will not attack the resist. Next, the resist is given a more substantial postbake, followed by another cleaning of the non-resist areas. Now the glass can be etched, as outlined above.

Another alternative for deep-etching is to ball-mill some aluminum stearate with the photoresist for several hours, and then coat the glass substrate with this mixture (Kodak Metal-Etch Resist Additive D Formula). The aluminum stearate thickens the photoresist and increases both its adhesiveness and its ability to repel acid. This formula can also be used for photoetching ceramic surfaces.

b. In *shallow-etching*, heat-polymerizing epoxy ink or an intermediate barrier of silver, as used in mirror fabrication, can be employed. A similar procedure to that outlined in *a.* is recommended. When working with silver, however, ferric chloride or a weak nitric acid solution must be used to remove the silver from the resist-free areas. With care, flashed glass can be shallow-etched quite satisfactorily without an intermediate layer.

3. Hyalographic transfer technique: This third way to create the acid-resistant image is a version of the dusting-on process (detailed under nonsilver processes in Chapter 7). A positive transparency is used to create a latent image on a piece of paper that has been coated with a solution of gum arabic, sugar, ammonium dichromate (6 grams of each), and distilled water (60ml). The image is "developed" in this instance with aquatinting powder. When the powdered image is brought into contact with a piece of heated glass, the powder particles transfer. The vapors of hydrofluoric acid are generally recommended as the etchant.

B. The Use of Photosilkscreened Matte Etch, Hydrofluoric Acid Compounds

The front windshields of most cars are made with safety glass, verified by close inspection of the little white marking on the window. This matte-textured designation was made with a glass-etch compound. Matte etch solutions were designed for industrial purposes, but they also have tremendous artistic potential. The etchant is packaged in a working consistency for silkscreening, although

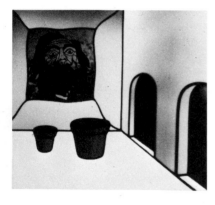

Paul Marioni, "The Witness." The image is a high contrast positive transparency sandwiched between a sheet of clear and a sheet of blue opal glass. Image was then integrated into the stain glass window. 24" x 26".

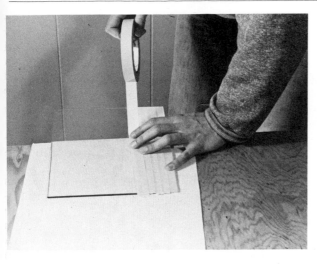 

images have been made successfully on photoresist-prepared glass by dipping, using this consistency. Some companies (e.g., McKay Chemical Company) market a thinner for altering the consistency of the etch fluid. It is normally not recommended to use more than 10%, by weight, of thinner in relation to etch, as it will alter the frosting ability of the etchant.

When using matte etchants, it is best to use a negative halftone or line transparency to form the image. Opaque the background of the transparency where necessary to have the glass frost in that area. If the original subject was photographed against a plain white background, then the enlarged transparency will only require minor opaquing. With clear window glass as the substratum, a frosted ground with a transparent image is most pleasing. The most suitable images to etch in glass are those that produce fine lines and small expanses of clear, unetched glass.

Matte etchants tend to crystallize when cold. Therefore, before using one heat it in a tray of hot water to rid it of any lumpy formations. As an added precaution, it is also advisable to strain the compound through cheesecloth or a piece of the screen material itself, ensuring a fine, smooth flow through the mesh. When working with matte etchants, wear rubber gloves, use only plastic trays, and make sure there is good ventilation. These etchants can be purchased from silkscreen and stained-glass suppliers.

Basically there are two methods for application of the image when using matte etchants;

1. Make a photosilkscreen of the image and screen the etchant. If photosilkscreen technique is employed, be sure that a water-based direct emulsion, like Encosol-3, is used. One should be aware that the emulsion does tend to break down, causing some loss of image detail after about 20 screenings of the compound. This occurs even though the resist is water-based. It is due to the hydrofluoric acid base and to the slight abrasive quality of the matte etch compound itself. Because of the amount of set-up time involved, photosilkscreening is best reserved for multiple pieces and the production of posterized images.

Photosilkscreen Matte Etchant Procedure. 1. (above left). Clean printing surface with extra care and tape backside of glass. 2. Warm etchant to remove crystalline formations; load image area with etchant using squeege; insert glass; squeegee image using basic photosilkscreen technique.

Gail Griggs & Roger Ostrom, "Elevator Tower Window." A photo-collage, sandblasted and mirrored, depicting a visual history of the local environment from an historical and contemporary perspective. A sandblast resist was developed and photosilkscreened on the glass. Each window pane is approximately 3' x 4'. Total size of the piece is 8' x 32'.

Before pulling an image on a good sheet of glass, pull a trial print on a waste piece of glass or on paper. This preloads the mesh with etch and thus helps to insure that the image will etch evenly. Avoid making more than one pass with the squeegee. Also be sure to tape the edges and back of the glass, as well as the piece itself, down to the printing board. The latter precaution will keep the glass from sticking to the underside of the screen and smearing the image as you pull the squeegee across the mesh. (More on silkscreening in Chapter 9.)

These etchants will frost glass in about two minutes, although there is an exceptionally large latitude in the amount of time necessary to acquire an acceptable matte white tone. After the etchant has remained on the glass for a few minutes, place the glass under running water and wash off the compound. Next, place the piece in a vertical position and air dry it. This is a very simple process, the image being permanently integrated into the glass.

Photosilkscreened glass etchants can also be used to create tones in the image. It has been established that rescreening or re-etching the glass with these compounds will alter the tone, the degree of translucency. This is especially useful in the screening of simple posterized images, where up to four transparencies of the different densities are made from the same image and used to make a semi-continuous-tone reproduction. Screen the image made from the first transparency; leave the etching compound on the glass for approximately 30 seconds and then wash it off with water. Align the screen made from the second transparency carefully with the image left by the first one and screen it, leaving the compound on this time for one minute. When screening the third transparency, leave the compound on the glass for two minutes, and when working with a fourth, leave the compound on 15 minutes before removing it. Each additional screening creates a new, deeper tonal value in the image by re-etching parts of the glass.

2. Use a photoresist to create an image on the glass, dip the glass, or screen on the etchant. It is preferable to use a photoresist system for one-of-a-kind pieces. The use of a photoresist also makes it easier to etch images on surfaces that are not flat.

No intermediate adhesion layer is necessary. Begin by following normal photoresist procedure. It is very, very important that the surface of the glass be completely clean and free of grease. The slightest oil from your fingers may cause blotchy results. Clean the glass with whiting and household ammonia until a stream of running water will flow unbroken over the surface. Acetone is also a good cleansing agent. Dry the resist thoroughly after application and bake after development (say, five minutes at 93° C; 200° F). Always heat glass very slowly to avoid possible cracking.

a. To dip-etch the piece, fill a plastic dish with the etchant and immerse the glass face down, quickly but steadily in one continuous motion. (Of course you are wearing protective apron, gloves, and goggles.) Be careful when you dip the glass; the etchant starts reacting immediately. If a slppy coating is applied, the glass will have streaky and spotty tones. Next, withdraw it slowly, making sure the entire surface has been covered, and allow some of the excess to run off. Place the piece in a horizontal position, face up, on a flat surface and let it lie there for two minutes. It is important that the coating of etchant be reasonably consistent in thickness.

When the time is up, wash the glass in running water to remove the residual etchant. Matte-etched glass seems to attract unsightly fingermarks, but these are quickly removed with window cleaner or hot, soapy water.

b. The alternative method for applying etchant is to use a blank silkscreen. Mask off on the screen an area the size of the glass, using masking tape and butcher paper. Then simply squeegee the etch onto the glass through the screen. This method insures an even matte tone.

Use the means of application best suited to you. The secret in using these glass etchants is to apply the compound evenly.

Experience indicates that a commercial photoresist tends to leave a slight oily film after development. At this point in the process, it is very difficult to reclean the glass without damaging the image. Although these resists produce a finer halftone with less bleeding of the etchant than in a photosilkscreened image, a blotchy background will be evident on the glass. To minimize this problem, yet still yield a superior halftone image (65-line) when using matte etchants, use the homemade resist detailed under photo-sandblast-resists in the next section. Just before etching, dip the glass in a very mild, soapy, hot-water bath. This will help to clean the open image areas. Rinse with cold running water. Once the glass is completely dry again, use alcohol and cotton balls to clean up any lingering grime. The glass should now be clean enough to etch. To deepen the matte tone, rinse off the etchant in water and let the resist air dry, then rescreen or redip the piece. Be careful not to handle the resist while it is wet.

A detail of "Elevator Tower Window."

ADDING COLOR TO MATTE-ETCHED GLASS

It is possible to add color to glass either before or after etching. This may be done in several ways. One might try photosilkscreening a durable, yet transparent, mineral-based paint onto the glass in the form of either a negative or a positive image. Mineral-based epoxy paints can also be used as a resist, thus making possible the creation of a frosted image and its colored component in fewer stages. Oil-based images will not be as permanent as those etched into the glass, but one might lessen damage to the painted surface by laying another sheet of single-strength glass over it; the two sheets could be held tightly together with lead came around the edges (available from stained-glass suppliers). Other possibilities might include photosilkscreening the etching fluid, onto a sheet of colored glass or overlaying etched glass with a piece of colored acetate.

SANDBLASTING PHOTO IMAGES ON GLASS

The technique of sandblasting a photographic image into glass is relatively new, although hand-cut sandblast-resistant stencils have been used for years to create patterns on glass. The following data deal with a homemade photoresist used to create sandblasted imagery on glass. Although only glass is discussed, the resist and the technique are readily adaptable to other materials that can be etched with abrasives and, in certain instances, acids. Alabaster (a soft stone), animal bone and ivory, wood, etc., are all worthy of experimentation.

Sandblasted glass creates an interesting crystalline effect when light passes through it. This is more noticeable close up than from a distance and is the major visual difference when compared to a matte-etched piece of glass. The real benefit of sandblasting is that it is much less hazardous than acid. It is also easier to employ on the larger surfaces, granting the user better control over the evenness of image tonality. In addition, when sandblasting is done on flashed glass with two-color flashing, there is a potential for increasing the variety of hue by altering the translucence of one or both colors.

Light-Sensitive Sandblast-Resistant Materials

A resist must have certain qualities if it is to be used on glass that will be sandblasted. It must either be tougher (more abrasion-resistant) than the glass or else so rubbery and flexible that the sand particles will be repelled by it. It must also be capable of both bonding to the glass and being easily removed. Since the resist is to be employed in image-making, it must be either light-sensitive or capable of being silkscreened. The following photoresist exhibits these qualities, enabling one to reproduce images photographically with the aid of a sandblast system:

- 1 Tbsp of polyvinyl-alcohol/polyvinyl-acetate, the base component (an unsensitized, medium viscosity direct photosilkscreen emulsion may be used as a substitute)
- ¾ tsp of Elmer's School Glue
- 3.8cm (1½ in.) worm of Translucent Phenoseal (a water-based plastic vinyl substance, made by the Gloucester Company, that bonds well to glass surfaces and is permanently flexible, yet tough)
- ¾ tsp of ammonium dichromate sensitizer (stock solution: 1 part ammonium dichromate to 5 parts water)

Materials and chemicals for photo sandblast resist application on glass surfaces: ammonium dichromate and additives; size for size transparency; direct photo emulsion base: waterspray bottle.

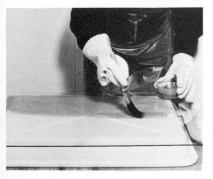
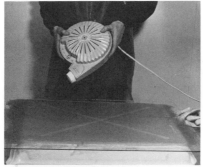
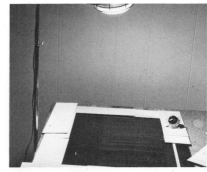
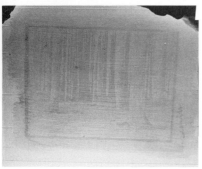

● 1 Tbsp of water (water is added for thinning; a heavy viscosity direct screen emulsion can be added for thickening).

A diazo-type direct photosilkscreen emulsion may be substituted as a sensitized base solution, although exposures will be longer. Phenoseal, however, should not be included when glass is being precoated for exposure at a later date. Generally speaking the additives (Phenoseal and glue) help to eliminate annoying bubbles, strengthen the resist coating, and improve the adhesive properties of the emulsion.

To make a working solution, mix the emulsion base with the two additives. Add the ammonium dichromate solution to sensitize the mixture just prior to use. Thin or thicken per requirements at this time. The above quantity of resist will coat two 20x25cm (8x10 in.) pieces of glass. All steps from coating through development of the image should preferably be done under safelight conditions.

Apply the emulsion with a brush or squeegee it through a blank silkscreen. If you desire, the emulsion can be screened, while still unsensitized, through a photostencil. The important thing is to keep the coats thin and even, to minimize drying problems such as hairline cracks in the emulsion's surface. Dry between applications with the aid of a hair dryer. For halftone images (40-line to 65-line) apply two very thin coats. Where you need maximum sandblast resistance (e.g., line images), however, use an undiluted or thickened mix.

Soon after the resist dries it should be exposed to UV light, which hardens the emulsion corresponding to the clear areas of the transparency. Lay a negative transparency, emulsion side down, on the emulsion of the coated glass and place a sheet of clear glass on top. Put a piece of black cardboard under the coated glass to absorb stray

Photo Sandblast Resist Technique (left to right from top). 1. Coat the sheet of glass with the resist in subdued light. 2. Dry the resist coating. 3. Expose the sensitized glass/transparency sandwich to an ultraviolet light. 4. Latent image after exposure and prior to washout. 5. Wash out the unexposed, unhardened resist areas. Use a spray bottle for clearing areas of intricate detail. Dry for 24 to 48 hours and sandblast the glass. 6. The final sandblasted image on glass.

Above top: A halftone image sandblasted on glass. Notice the longer tonal scale—the dot pattern is about 50 line.

Above: the glass need not be cut in a rectangle to a standard size in order for the image to be effective.

Right: A cut out, collaged image of several tonal separations in order to hold detail. This is an alternative to the use of a half-tone screen.

Kent E. Wade, Untitled Images

light. Exposures will vary with the quality of the light source, the thickness of the coating, and the distance between the UV source and the emulsion. Exposures of 20 minutes to 45 minutes with a sun lamp at approximately 63.5cm (25 in.) are common. Shorter exposures are attainable, and slightly thicker resist coatings are permissible, with the use of a photo-type arc lamp.

Prior to sandblasting, paper stencils can be cut out and applied if you want to mask off any image segments (suc as distracting backgrounds). It is amazing how well even a piece of ordinary masking tape holds up to the blasting, when compared with other resist materials. Remember, too, that prior to exposure of the resist, distracting details can be painted out with photographic opaque on the transparency itself.

Although most substances will wear down eventually with repeated blasting and abrasion, metal can withstand such abuse better; therefore, as an alternative resist material, in particular for duplicate glass images, consider making thin, 24-gauge, metal stencils. They can be created by photoetching completely through the metal (review metal throughplate etching) and then gluing the stencil directly to the glass. This technique, however, would be limited to

After the transparency-emulsion sandwich has been exposed, "develop" the image immediately by flushing the surface with hot water (43° C; 110° F) to remove the unhardened areas of the resist. A small brush and a plant-misting device are useful for this. The resist emulsion is fragile when wet; so avoid handling the surface. After development, and once the image itself is dry, surface alterations can be made: One can paint in some areas with resist, touch up or remove other areas. To remove resist, cut into it with a mat knife or razor, rewet just that area, and carefully scrape off the unwanted portion. Now set the glass aside for 24 hours to cure the Phenoseal.

the kind of images that could be etched successfully in metal. Experimentation might begin with thin lead-foil sheeting, copper foil, or used aluminum offset plates.

Sandblasting the Glass

Equipment for sandblasting can be very expensive. One might, however, procure the services of a local sandblasting company. It is best to work with one individual, who can become familiar with your requirements. Use about 40# to 60# of pressure and proceed slowly. Sandblasting should be done perpendicular to the surface of the glass, so that the grit is less likely to undercut at an angle and blow the resist off.

Different grit sizes create different visual effects on the glass; thus one has a veritable palette from which to choose. Sand and silica-carbide grits can both be retrieved and used again; so try to keep the different grit sizes separated when blasting. Use a cardboard box large enough to hold the glass, and cover the opening with clear plastic sheeting. A hole in the plastic will permit the glass to be blasted, while confining the grit. Always wear goggles and a mask when blasting. Be careful not to breathe in the very fine particles that will float about the work area.

For general-purpose work, use an 80-grit sand or a 60-grit sand (recycled). As the sand is used, its edges become rounded and its particle size diminishes. Glass appears to be blasted only so far with one size of grit. When a very fine, flat finish is wanted in part of the image, switch to a finer grit (120#). A 120 grit gives a finish closer to that of a matte etchant. If a more crystalline effect is desired, then switch to a coarser particle size.

On glass, a frosted background tends to look better. Therefore, photograph subjects against a white background, so that that area will be opaque in the negative transparency and thus shield the photoresist emulsion from hardening under the ultraviolet light. When the photoresist is processed, the unexposed emulsion, representing the background, will wash away baring the glass there to the abrasive action of the sand. Monitor sandblasting progress continually until the desired degree of frost is attained.

If you want to render extremely fine detail, or halftone images, experiment with the variation that follows as a means for improving final visual results. Once the image has been created in the resist layer, touched up or further manipulated, and dried again, apply a very light coating of matte etch and process normally, but do not remove the resist. Next, blast the glass with low pressure and a very fine grit in order to alter the white matte effect slightly. Minimal blasting will be required to obtain a pleasing halftone, with less chance of blasting off the resist dot pattern (see Chapter 8 on dots in halftone transparencies). Lastly, the main subject area can be masked and other image segments blasted with a coarser grit.

Once the glass has been sandblasted, remove the resist by soaking the piece in a bath of hot water. Household bleach added to the water will help expedite resist removal.

PHOTOSILKSCREENING IMAGES ON GLASS

Photosilkscreening is a useful tool for transferring an image to a glass surface either directly, via special glass paints, or indirectly, via decals. Some of the paints require firing in a kiln; others do not. Let's examine some of the options.

1. Nonfired images. Glass paints that do not require firing naturally do not require a kiln. These paints provide a certain amount of permanency, some more than others. The best sources for products and technical information are local paint stores, stained-glass dealers, and screen-printing supply houses. A company that markets glass paint specifically designed for screening is the Advance Process Supply Company. Refer to their FED Series of fast-drying high-gloss enamel inks. This paint should be used with a reasonably fine mesh in order to bring out optimum detail. It is most important that the glass be very clean prior to paint application. As this paint remains on the surface of the glass, it is advisable to consider some protection for that surface during its display.

2. Fired images. Certain glass paints must be fired. Here, the paint (made with finely ground particles of colored glass) is permanently fused into the glass substratum in the shape of a photo image.

As an alternative to silkscreening these paints, a direct emulsion base like the one used for photoenameling (Chapter 1) or for the direct photoceramic process (Chapter 3) could be tried. Here, the base emulsion is loaded with the appropriate glass pigment and baked. The base component fires out free of ash at a low temperature, leaving the colored frit to fuse onto the glass sheet as the proper firing temperature is attained.

The dusting-on process also offers another interesting alternative (see nonsilver processes, Chapter 7).

Technical information regarding firing technique, kiln requirements, and paint preparation information can be obtained from any good book on stained-glass production or directly from glass paint manufacturers. Generally the firing range for these paints is between 530° C and 567° C (1000°-1050° F). It is important when firing to let the glass cool slowly in the kiln. Some people suggest firing by observation: Once the paint has a high sheen to it, it is time to start the slow cooling cycle.

Several companies sell glass paints that can be screened and fired:

a. The Standard Ceramics Supply Company markets a complete range of glass colors that can be screened by mixing with their heavy medium (No. 721) to a pastelike consistency. These glass colors fire between 566° C and 626° C (1040°-1140° F). The company recommends starting with three parts, by weight, of dry color to one part of No. 721. Although the correct consistency of the mix will vary with the mesh size of the silkscreen (a monofilament polyester is a good choice), this ratio is suggested as a general guide. Standard Ceramics also sells ceramic inks. It is not advisable, however, to use a ceramic ink on glass. Ceramic inks are normally employed as an overglaze color, the firing temperatures being higher than for glass paints.

b. L. Reusch & Co. markets several products for use on glass, such as a glass-etch compound, Drakenfeld glass colors, and glass-decal screening mediums and covercoats. Each color fires differently and each offers its own set of problems. Black colors are the easiest to

use. Try #24-114 Black Drakenfeld Soft A.R. Color. Mix to a screening consistency by adding 25% gum arabic to 75% pigment plus water. As a substitute vehicle for the gum, a decal medium, a squeegee oil, or even linseed oil slightly thinned with turpentine can also be used (vegetable-oil-based mediums are to be avoided). Screen the image and let the glass dry thoroughly before firing the piece. Apply kiln wash powder to the shelves of the kiln, then lay the piece of glass down in a horizontal position. With the door slightly ajar, fire the kiln up to 482° C (900° F), then close the door. The color matures between 582° C and 593° C (1080°-1100° F). If you are interested in slumping the glass over a clay bisque mold, then fire on up to 654° C (1250° F) and turn off the kiln. Let the glass cool overnight with the kiln door closed, to prevent thermoshock. The vehicle fires out in the kiln. As an added precaution, apply kiln wash over the mold to act as a release agent.

    c. Other colors should also be tested. The Drakenfeld colors are the most colorful; however, they are reflective, not transmitting colors and will appear dull or gray if light passes through them. On the other hand, glass-stainers' colors, which are fired higher, are normally viewed with the light behind them (transmitted light), but their color range is limited in comparison with Drakenfeld's. There are also transparent silver-stain colors for glass. These colors react with the flux in the glass to form yellow and orange hues. Transparent colors offer another means of presenting posterized imagery, either by superimposing and firing one color atop another or by superimposing several sheets of glass, each containing one image component, sandwiched together and backlit.

    d. Finally, there are several low-firing silkscreen inks that are really epoxy resins, which dry by polymerization. These inks re-

*Photosilkscreened & Kiln-fired Images on Glass (left to right from top). **1**. Materials required: glass pigment mix; printing vehicle, glass cutter, squeegee, photosilkscreen. **2**. Squeegee the image onto glass substrate. **3**. The screened sheet of glass. **4**. Cut the images into manageable sizes for kiln firing. **5**. Liberally apply kiln wash to shelves of kiln. **6**. Kiln fire to permanently fuse images to glass.*

quire a catalyst and are heat-cured to increase their adhesive properties. (Curing temperature range is 65°-260° C; 150°-500° F.) In addition, they do not melt and fuse into the glass, but rather bond to the glass surface. In this respect they are similar to a nonfired ink, except that the bonding is a great deal stronger. The Naz-Dar Company markets several such inks. The Atlas Screen Printing Supply Company also markets similar products, such as the Wornow Cat-L-Ink 50-000 Series. (Use it with Baking Catalyst #45 for superior results on glass.) This ink is available in the following colors: yellow, orange, green, blue, red, brown, black, white, and clear.

SPECIAL GLASS PLATES AND PHOTOEMULSIONS

If you do not wish to employ photoetching, silkscreening, or sandblasting techniques, it is still possible to create images on glass by utilizing special photosensitive glass plates or photoemulsions coated on the surface of the glass. Eastman Kodak manufactures several different glass plates for scientific applications. Although these plates are readily adaptable to any creative endeavor, they can be expensive. If interested, however, inquire about one of the following types: **1.** Kodak Ortho PFO Plates. These have a similar emulsion to Ortho Film Type 3. They can be used under a red safelight and processed with Kodalith developer. **2.** Kodak Photoplast Plates. These are similar to PFO plates but, instead of a glass support, they have a clear acrylic plastic base that can be drilled or cut into any desired shape. **3.** Kodak Aerographic Positive Plates. These come in either a medium or a contrasty emulsion and have an antiabrasion covercoat. They can be used under red safelight and processed in Dektol developer. All these plates can be ordered through a Kodak graphic-arts dealer.

A piece of glass can also be coated or sprayed with a liquid emulsion. The emulsion can be made at home, or a commercially prepared brand can be used. The Rockland Colloid Company recommends BB-201 (a fast, enlargement-speed emulsion) for glass. This emulsion has extra density, which assists image visibility on transparent materials. A special subbing powder that will increase the adhesion of the emulsion to the glass surface is included with the Rockland emulsion kit. Generally speaking, this emulsion acts like that of a black-and-white photographic paper and must be used under darkroom safelight conditions. Just follow the explicit, easy directions. (It can be used on many other surfaces, as well. Review uses of Rockland emulsions detailed under photoceramic processes, Chapter 3.)

Norland Products, Inc., markets an emulsion called Norland Glue-Silver Emulsion, NGS II. It consists of a silver chloride solution suspended in a fish-gelatin colloid. To use the emulsion, the base colloid is sensitized with ammonium dichromate and coated onto a clean glass surface. A transparency is placed in contact with the coated glass and both exposed to ultraviolet light. The light that is able to penetrate the transparency will harden the emulsion. During water development the unexposed gelatin-silver emulsion will wash out, leaving a relief image on the glass plate. Kodak Dektol paper developer is next employed to develop the remaining light-sensitive silver left on the glass, turning the relief image dense black. This emulsion can also be used on other surfaces. (See the section on dichromated colloids in Chapter 7.)

PRESENTATION OF GLASS IMAGES
There are a multitude of ways to enhance image presentation on glass surfaces. Here follow a few ideas.

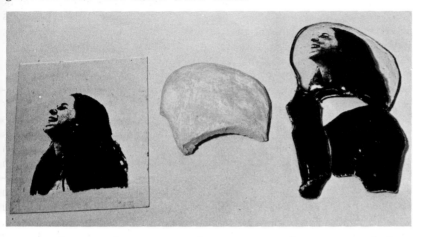

A photo-sculpture begins to take shape with pieces of stained glass added using copper foil technique. (from left to right): Image photosilksceened on glass and Bisque mold over which glass is slumped in kiln. See color section #1 for finished piece.

1. A photographic image can be incorporated into a stained-glass window. For example, using lead ribbon (called "came") and different colors of glass, a seascape design might include the suggestion of a bird: A photographic image of a real bird integrated into the glasswork will accentuate that element of reality.

2. On the other hand, an image on glass, perhaps a portrait, might be given an unobtrusive lead frame to accentuate just the image itself. A series of etched-glass images could be constructed this way, suspending them one beneath the other from copper rings soldered to the lead came.

3. A viewing box can be created for image presentation, perhaps aligning several images in a row with space between them and a light source at the rear of the box for illumination. A fully frosted piece of glass close to the light source would help diffuse its harshness. The series might consist of one subject at various stages of motion. Viewing the image from different angles than perpendicular to the front of the box would alter one's perspective, adding to the visual enjoyment of the piece.

4. Images can be etched in glass and then filled in with other substances (paint, enamel etc.).

5. The silver layer of a mirror can be photoetched. Mirrors are well suited to the presentation of some images. The reflective quality of a mirror, its ability to add depth and light to a room, is an attribute that should be considered in image selection. Most mirrors consist of three or four layers of material. First there is a black protective coating that normally can be removed with lacquer thinner. Next there is a thin copper layer, superimposed on a metallic silver layer that is coated onto a sheet of clear glass. After removing the backing material, clean the copper-silver sandwich very gently

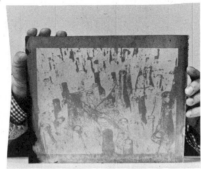

Photoetching a Mirror (left to right from top). **1.** *Remove protective backing with lacquer thinner.* **2.** *Apply a coat of photo resist on the cleaned copper/silver surface.* **3.** *Dry the coating and expose the transparency/resist sandwich to an UV light.* **4.** *Develop image for 2 minutes in the resist's companion developer.* **5.** *Photo resist image prior to etching.* **6.** *Etch image with ferric chloride until all non-resist covered areas are removed.* **7.** *The mirrored image as seen from the front.*

and coat with a photoresist. Lay a transparency, emulsion side up, on the resist. Remember that the image is being exposed from behind the glass, but it will be viewed from the front. Thus, the image must be reversed at the time of exposure. Make the exposure, develop the resist in its companion developer, and postbake (optional). Next, etch the image in a bath of ferric chloride (review photoetching of metal surfaces, Chapter 1). The image will be seen either in metallic silver with clear glass for background or vice versa, depending upon whether one used a positive or a negative transparency. The image may now be backed with acetate or paint, adding color to the nonsilver areas as well as restoring a protective backing to the piece. Also consider electroforming images on the surface of glass, using silver as well as other metals.

6. Images can easily be montaged and sandwiched between layers of glass. Some materials worth exploring are hand-tinted black-and-white transparencies, enlarged full-color transparencies, colored indirect photosilkscreen stencil film (applied wet to the glass surface and hardened), color overlay films, and holographic images on acetate film.

7. Finally, any glass object or surface can be photoetched, photosilkscreened, or coated with a photoemulsion, and there is the added possibility of combining several techniques.

Above: Naomi Savage, "The Chase." Photoengraving on copper, deeply etched, oxidized, silver plated and lacquered. 11" x 14"
Left: Larry Bullis, "Guckian's Fast Freight." Photo-fabricated copper piece, brass plated and colored blue with hypo/lead acetate solution using photo resist technique. The frame is brass. The image is an assemblage of several transparencies. 5" x 6-1/2"

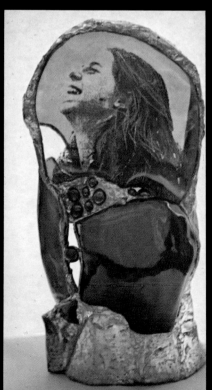

Facing page, clockwise from upper left: Eleanor Moty, "Reflection Pin." Photoetched silver with abalone shell and leather. Silver is colored (iridescent) with sulfur solution. Larry Bullis, "Peacock." Basic metal is brass. Image is etched through nickel plating. Blue color is applied to heated plate with hypo/lead acetate solution. Butterfly is photoetched into copper frame. 3-3/8" x 4-1/4" Eleanor Moty, "Landscape Handbag." Image is projection printed, silver electroplated rather than etched. Object is sterling silver, brass and agate. Martha Forster Banyas, "John's Box." Photosilkscreened brass lid using underglaze "D" enamel. Lid has been domed, fired, further enameled and ceramic decals applied.

Martha Foster Banyas, "Bruce's Box." Deeply photoetched copper via silkscreened resist method; enameled using Champlevé technique. Box fabricated from redwood.
Above: Gail Schoelz, "Chess Set." Fired-on photo silkscreened glass enamel, including stained glass, lead, copper foil and wood.
Left: Gail Schoelz, "Sarah." 3-Dimensional glass sculpture with fired-on, photo silkscreened enamel image on clear glass, slumped. Base is made of copper sheeting, leaded over with glass jewels and German stained glass integrated into it. A separate piece of colored glass is slumped behind the image giving the image color and the piece itself added depth. 18"H x 8"W x 7"D

Kent E. Wade (photography), Lynn M. Weckert (stained glass),
"Miller Family Portrait." A 65-line halftone image created using a
photo sandblast resist. The image was matte etched on double
weight clear glass using two separate silk screenings of Screen Etch
to deepen the tone. A stained glass window frame was designed to
complement the image.

Above: Richard Posner, "The Big Enchilada." Flat glass panel with positive Kodalith and Cibachrome transparencies integrated into the stained glass picture window. Images each sandwiched between two pieces of clear glass and sealed at the edges via copper foil technique. 33" x 45"

Above right: Paul Marioni. "25 Years." Stained glass piece done to commemorate the couple's 25th wedding anniversary. Cibachrome transparencies were made from 35mm slides of the couple, enlarged to fit the full-size drawings. Transparencies are sandwiched between two pieces of clear glass. 28" x 27"

Right: Paul, Marioni, "Plug for Those Who Dare." Stained glass. A pictorial memory of how fast one can take that turn. Photo image of hairpin turn in Cibachrome transparency sandwiched between clear glass, while Chrysler Airflow is photo silkscreened glass enamel baked on clear and white glass. 24" x 24"

Right: Mary Ann Johns, "Funny You Don't Look 32," #2. Rockland Print-E-Mulsion was applied to underglazed bisqueware. The processed image was then stained with a water-based acrylic. Details were applied by hand using watercolors. A polymer coat protects the artist's self-portrait from abrasion. 2   4' x 3"

Below: Jugo de Vegetales, "Diver." Life-size photo ceramic decal on whiteware. Airbrush employed to create illusion of depth.

Below right: Les Lawrence, "Vase, Made-in-U.S.A." Images are photo silkscreen transfers from shoebox tissue paper using a black glaze stain. Multiple images are transferred to leather-hard clay which is then reshaped, airbrushed and painted wi   erglaze colors. The cupids are made via press mold stamps and th   piece is finally bisqued fired, sprayed with clear glaze, other overglaze pigments and refired.

Left: Les Lawrence, "Plate, Made in U.S.A." Fabrication process is the same as vase at lower left, previous page.
Below left: Kent E. Wade, "Mother's Day Pillow." Light sensitive Inkodyes, employed on cotton fabric, stuffed and trimmed with frill.
Below: Lori Solondz, "Oh Madonna." Handcolored Cyanotype image on fabric using watercolor paints with a full-sized transparency laid over, stuffed and sewn together. 28″ x 8″ x 6″

Left, top to bottom: Lori Solondz. Untitled. Muslin fabric soaked in Van Dyke sensitizer. Image handworked with watercolor paints, sewn & stuffed. 18″ x 24″

Catherine Jansen, "Life-size Soft Bathroom." 3-M Color-in-Color images transferred to fabric including blueprint images on cloth.

Diane Sheehan, "Lovers Luncheon." A disperse dye image transfer on polyester fabric using B & W photostatic technique with 3-M Color Layout Sheets. Pieces were collaged and sewn. 13″ x 20″

Above: Mary Ann Johns, "A Quilt is a Quilt is a Quilt...." A piece on muslin, first dyed, then Rockland Print-E-Mulsion applied, handpainted with fabric

then sewn and embroidered. 5′ x 5′ x 2″

Above: Catherine Jansen, "Photographic Pillow." Collage using blueprint technique and Inkodye Red on cotton fabric embroidered, sewn and stuffed.

Photoceramic Processess

There are many different methods for affixing a photographic image to a ceramic surface. Some processes incorporate photosilkscreening technique, while others rely on the direct application of a light-sensitive, glaze-based pigmented colloid to the surface. With most of these processes the photo image is permanently fused or bonded with the ceramic piece during kiln-firing. In addition, there is some choice as to the stage at which one may incorporate the image. For example, the image may be integrated into the ceramic piece while the clay body is still moist, after bisque-firing, as an underglaze or an overglaze, etc. Not only the time of image application, but also the technique utilized, will affect the final result. Through ongoing experimentation with these techniques, one will begin to establish methods of application that will best suit your personal needs. The following is an examination of some of the more interesting photo-ceramic processes.

METAL PLATE IMPRESSIONS IN CLAY

Any photoetched metal plate can be utilized to impress a photo image in clay. Of course, the deeper the bite, the better the image transfer. As in other transfer techniques, the image will reverse. This process is similar to pressing animal cookies in dough with metal molds. Images that recess into the plate will be raised on the clay, while images that are raised in relief on the plate will be recessed (intaglio) in the clay body. The plate is pressed against a piece of leather-hard clay. Talc powder sprinkled over the clay surface helps to prevent the clay from adhering to the plate as it is removed. A thin coating of all-purpose oil may also be used to accomplish the same end. Once the image has been transferred to the clay slab, it may then be molded into any desired shape or incorporated into a larger piece prior to bisque-firing. After the piece has been fired, selectively brush a ceramic stain (e.g., iron oxide) into the recesses. This helps to delineate the image and improve its contrast with the clay body.

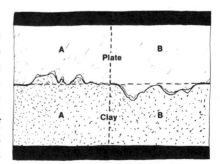

A. Plate photoetched Intaglio fashion creates a relief effect on the clay surface. B. Plate photoetched with a relief image creates an intaglio (recessed) impression in the clay body.

IMAGES OF SLIP OR STAIN

This process requires the use of photosilkscreening technique. Here, the photo image is transferred to a leather-hard clay slab by silk-screening on a glaze-treated slip or engobe. It is important that the slip be compatible with the clay body. A slip made from the original clay body is recommended, to avoid shrinkage problems. Once the image has been screened, the slab can be incorporated or shaped into the final piece.

Prepare a screen for either a halftone or a line image. It is recommended that one employ a water-resistant direct emulsion, such as Encosol-3, and a 157-to-200 mesh monofilament polyester or a stainless steel mesh. Experimentation is encouraged in the employment of a variety of mesh sizes, although the finer the mesh count, the less the quantity of slip that will be deposited. Any problem with the slip's drying in the screen can be eliminated easily with water without damaging the photostencil; however, it is important to keep the screen dry when it is being used. Any water on the fabric will cause the image to run because the consistency of the slip will be altered slightly.

One may use a variety of metal oxides or underglaze stains to give the slip color. If necessary, all ingredients should first be ground or ball-milled to about a 400 mesh. It is very important that the engobe pass easily through the screen mesh with minimal abrasiveness. Manufacturers' specifications indicate that the particle sizes of most ceramic ingredients vary substantially (e.g., cobalt oxide—200

"Medicine Shield" by Marilyn Higginson. A photosilkscreened underglaze stain and white engobe—10" diameter clay slab with feathers.

Photosilkscreened images using underglaze stain. (left to right from top) **1.** Coat the screen with direct photoemulsion in subdued light. **2.** Dry the surface with heat and/or fan. **3.** Expose the emulsion. Note that the screen is being exposed with the transparencies laying on the inside of the screen. **4.** Wash out the unexposed image areas after exposure. **5.** Prepare clay slabs to accept the image. **6.** Screen the image. **7.** The clay slab has not been bisque fired so it can still be manipulated, if desired. After the clay dries out, it may be fired in a kiln. **8.** Clean out the screen immediately after use.

mesh; iron oxide—325 to 400 mesh). Be liberal in the addition of the coloring agent to the mix, thus insuring good image contrast. The final slip for screening is prepared by the addition of water until a pastelike consistency is obtained.

Before screening, it may be advisable to lay down a white pigmented underglaze to act as ground. This will strengthen image contrast, as will the use of a light-colored clay body. Once the white background has dried, the clay slab can be registered under the screen and the prepared slip squeegeed onto the piece. It is best to work on the slab while it rests on a piece of construction-grade sheetrock. This chalky substance will help to pull moisture out of the clay body prior to firing, as well as give one a means of moving the slab during process steps without damaging it.

A slight texture added to the clay slab will permit more slip to be deposited. Apply texture by laying a piece of canvaslike material over the slab, once the background color has been applied. After screening, gently lay thin, absorbent tissue paper over the image in order to protect it during further manipulation of the piece. The ceramic piece can now be bisque-fired and glazed. A transparent overglaze of thin consistency will insure that the clarity of the image is not impaired.

There are many variations to this technique of employing silk-screen to lay the image on greenware. An underglaze stain or a finely ground oxide can be mixed with a screening medium in order to transfer an image onto a light-colored clay body. Try mixing some iron oxide with Naz-Dar Binder Varnish #5549. Not all screening vehicles will work satisfactorily. Use the eyeball technique to test the opacity of the mix by rubbing a dab over a white card. If a thinner is required, try Naz-Dar Thinner/Retarder #5550.

The screened greenware slab is left to dry overnight to rid it of moisture. When the piece is kiln-fired, the vehicle will burn out, leaving only the image. At this point, additional color can be added or a transparent overglaze applied.

Another photographic transfer process for applying the image to a damp clay body utilizes an intermediate of tissue paper (shirt-box variety). Here one makes use of an underglaze stain and a clear screening vehicle to screen the image on the paper initially. The paper is placed in contact with a semimoist slab of clay, image side down, and gently rubbed with the hand to transfer the image. The paper is then removed. Several images can be applied to a piece by this technique, prior to firing. Images or portions thereof may also be removed, with the aid of the appropriate solvent, while the clay surface is still moist. Once the clay is dry, further manipulation of the image is limited to adding more underglaze color by hand or by airbrush. If one intends to reshape the slab, this should be done before the clay dries out. The piece is then bisque-fired, covered with a transparent glaze, and refired in an oxidation kiln.

NONFIRED PHOTOCERAMIC IMAGES

There may be instances where one will want to apply a photo image to an acquired ceramic piece. Here one can employ Rockland's Print-E-Mulsion or use a homemade photoemulsion. As with glass, one will have to pretreat any glazed ceramic surface in order to obtain optimum adhesion of the emulsion. Images will adhere directly onto bisque ware.

To use the emulsion, it must first be heated to approximately 36°C (98°F), as below this temperature the emulsion is in a solid gelatin state. Pour a usable amount into a glass dish under safe light conditions. Dilute the emulsion with warm water if needed, and coat the desired surface with a brush. Once the emulsion has dried, run a conventional photographic strip test of the chosen image on a scrap piece of white paper similarly sensitized. Process the test strip, using Kodak Dektol developer, stop bath, fixer, etc. Choose the best time, expose, and process the clay surface. The white emulsion base will turn transparent in the fixer, so that the image will appear in black contrasted to the color of the clay body or glaze.

Some Rockland emulsions are slow in speed when compared to

"Fantasy Foldout." A self-portrait by Mary Ann Johns. See pg. 58 for production notes. 3" x 12"

Using a photo emulsion on clay. (left to right from top) **1.** Heat the emulsion to 98°F. **2.** Coat the ceramic surface with the emulsion under safelight conditions. **3.** Dry the photoemulsion. **4.** Expose the emulsion via an enlarger. **5.** Develop in Dektol solution (1:2) with agitation (2 min.). A soft sponge or brush is useful on larger surfaces. **6.** Stop bath or water rinse (40 sec.) **7.** Fresh acid-hardening fixer with periodic agitation (8 min.) **8.** Wash in running water (25 min.) Hypo eliminator can be used to reduce washing time to 5 min. **9.** Redry the surface. **10.** Add any additional color or surface decoration.

"Confrontation with Illusion" #6 by Mary Ann Johns. 8" x 18" x 5". Rockland Print-E-Mulsion applied to unglazed bisqueware. The piece was stained with water-based acrilics, then handworked with watercolors. A protective polymere coating was applied once the piece had dried.

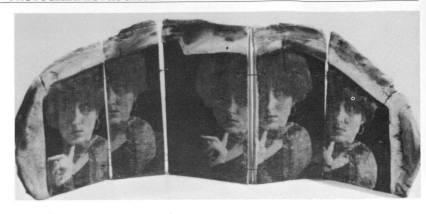

standard photgraphic paper. Rockland's CB-101, for example, requires an average enlarger exposure of several minutes. Images produced in this fashion are very susceptible to damage, as compared to images kiln-fired permanently onto the piece. One way to gain some protection for the image is to spray it with clear lacquer. Permanency may also be improved by low-firing the image, leaving a faint residual yellowish photographic stain (the image) on the ceramic surface after the gelatin base burns out. It is also feasible to convert the silver image by the mordanting technique, so that it is receptive to toners that are metal-oxide based. These could be low-fired for permanency and would afford some color variation.

PHOTOCERAMIC DECAL FABRICATION

Photo decal technique on ceramics is a transfer process whereby the photographic image is silkscreened with a finely ground china paint or overglaze pigment onto a special decal transfer paper. The decal is then positioned on a ceramic piece that has already been glazed and fired. After the decal is applied, the piece is refired.

It is an especially useful technique wherever curved or irregular

Basic equipment and materials for photoceramic decal fabrication.

ceramic surfaces are encountered; otherwise, a photosilkscreen system can be employed directly.

Decals can be made with underglaze stains for application to bisque ware; however, images of underglaze tend to bleed slightly into glazes applied over them. Thus, underglaze decals are more difficult to use than those made of overglaze pigment, which are applied to preglazed surfaces. Generally speaking, underglaze colors are more fully integrated into the piece, the image being protected with a clear glaze top. Overglaze pigments lie unprotected on the surface and will tend to scratch or wear off more rapidly, although even so one will find that overglaze decals are actually very durable.

Two silkscreens are required for making decals, one to screen the image and the other to apply the decal varnish or covercoat. Use a 200 mesh monofilament polyester fabric or a stainless steel mesh for screening the decal. This will enable both halftone images (65-line screen) and line images to be screened. A coarser mesh is preferable for application of the covercoat. A 10xx weave should prove satisfactory.

Oil-based pigments are employed for screening decals; therefore, use a diazo-type direct emulsion, such as Encosol-1, which

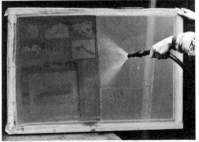

Decal Fabrication: (left to right from top) 1. Prepare all transparencies by cutting out unwanted background with a X-Acto knife on a light box. 2. Coat screen with direct photoemulson. 3. Expose transparencies to ultraviolet light source. 4. Wash out non-image areas and dry the screen. 5. Do any necessary touch up to emulsion with block out medium. 6. Pull images onto decal paper. 7. Images are dried then given a cover coat through a blank screen.

Decal Application: (left to right from top) 1. Materials necessary for decal application. Water and sponge, decals, glazed porcelain toothbrush holder and lusters and brushes. 2. Cut out the decal. 3. Soak decal in water. 4. Lay decal on ceramic surface and gently slide out paper backing. 5. Use a soft sponge or brush to assist removal of all entrapped bubbles. Thoroughly dry decal prior to firing. 6. While decal is drying, lusters may be applied. They can be fired with decal or in a separate firing.

permits the screen to be solvent-cleaned without damaging the photostencil. An indirect photofilm can also be employed. Try Ulano RX200, which is recommended for halftone work and for its resistance to abrasive inks.

The ceramic pigment utilized for decal work fires in the range of a Cone 017 to 019, oxidation kiln. There are several sources of supply offering a wide variety of overglaze colors. Choose colors that complement the glaze-fired background. In addition to the decal itself, lusters can be applied effectively around the image. Two companies are: L. Reusche & Co., who market Drakenfeld overglaze colors, liquid luster colors for china painting, and decal medium #1371; and Standard Ceramic Supply Company, who market overglaze colors for ceramics, lusters for glass and ceramics, and a heavy medium #721 for silkscreen applications.

Some companies, such as L. Reusche & Co., offer premixed overglaze pigment and vehicle in minimum bulk quantities. For premixed pigments packaged in small tubes and jars, one might try Versa Color. These are finely ground oil-based overglaze decorating colors prepared especially for screen printing on ceramic and enameled surfaces. Although not designed specifically for decorating glass, these pigments have been used successfully on glass surfaces. Versa Color is available in eight colors: yellow, red, orange, black, white, blue, green, and brown. Turpentine may be used to thin colors for airbrush application. These pigments are available from AMACO, the American Art Clay Company, Inc.

Ceramic pigment and vehicle may also be mixed in one's own studio. The vehicle, or medium, acts as a liquid carrier for the overglaze pigment. It imparts no color to the ceramic piece during firing. When mixing up pigment, be certain it has been sufficiently ground or ball-milled to minimize the oxide's abrasive characteristics. If one intends to use a 200 mesh screen, then the powdered pigment should be ground fine enough to pass at least 250 mesh to 300 mesh sieve. In addition, the pigment and the vehicle should be so blended that the mixture will flow easily through the screen mesh without clogging it. To mix a batch of pigment, add three parts of dry color to one part of medium. Add more or less vehicle for better screening consistency. Mesh size will ultimately determine the proper consistency required. Paint thinner or lacquer thinner can be used to unclog the mesh during screening and to clean up the screen at the end of a work session.

One will also need decal paper on which to screen the overglaze colors. A water-release decal paper can be obtained from Brittains (U.S.A.) Ltd., who market Thermaflat Decal paper, or from Atlas Screen Printing Supplies, who market Simplex Decal paper. Decal paper can be expensive, as it is normally sold in large quantities. Check with a local screen supply house to see if they carry individual sheets. If one feels adventurous and would like to make one's own release paper, experiment by coating a piece of soft drawing paper with a thin, even layer of gelatin, starch, or water-based glue. Once the paper is dried, it may be used like any machinemade transfer release paper.

(It is possible to purchase prepared one-color ceramic decals. Some companies will make decals from photographs submitted to them (H. Battjes & Co.). Premade multiple-colored decals are also available in a large variety of images. Thus, one does not necessarily have to become involved in lengthly decal fabrication techniques.)

Once the image has been screened onto the decal paper and dried, a protective covering is applied, also by screening. This is referred to as a covercoat, or varnish. It holds the ceramic pigment (the image) together during transfer. Covercoat may be obtained from Atlas Screen Printing Supplies (Wornow No. 25-810 Ceramic Clear Coat) or from L. Reusche & Co. (Drakenfeld Decal Covercoat, 3238). Apply the covercoat evenly, masking the screen so as to leave a thin border around the edges of the decal to facilitate handling while the paper is still wet. If one finds that the coating is too thin, use a coarser mesh to allow more covercoat to flow through the screen, or apply a second coating after the first one dries. Once the covercoat is dry, the decal should remain flexible enough to conform to almost any uneven surface. Dried decals can be stored between waxed paper sheeting until they are required.

A Step-by-Step Summation of Photoceramic Decal Fabrication

1. Prepare the silkscreen, using positive transparencies (see Chapter 9). Place the images so that several may be screened onto one sheet of decal paper at the same time.

2. Prepare the printing medium. In dry weather, add some boiled linseed oil and turpentine to squeegee oil in order to lessen cracking and lifting off of the image from the decal paper during the drying stage. One can also use a homemade printing vehicle composed of the above two additives plus some Damar varnish. A screening consistency is arrived at by using equal parts of each, mixing them into the pigment.

Toothbrush holder by Kent E. Wade.

3. Squeegee decals. A decal may be screened in multiple colors, using posterization or four-color separation techniques. Let the paint dry between coats.

4. Decals may be hung or laid flat to dry. A clothesline and pins make for a fine drying system. It will take from two to five hours for decals to dry, depending on the humidity. Forced drying with fans can reduce drying time.

5. Apply covercoat to dried decals in order to prevent bleeding of the image, then dry the covercoat and store those decals not needed immediately. Two methods are suggested for varnish application:

a. mask off a screen slightly smaller than the surface area of the decal paper and apply covercoat to the entire sheet;

b. make a nonphotographic, hand-cut stencil that permits covercoat to be squeegeed only through open image-related areas.

The latter method will save time normally spent cutting around each individual decal before its application. This technique is recommended especially where multiple sheets of one decal series are being made.

6. To prepare the decal for application, first trim off all the excess paper around the image, using scissors or an X-Acto knife for really fine detail.

7. Fill a pan with water at approximately 26°C (80°F) and place the decal face up in the water. After approximately 25 seconds, the decal should begin to curl and loosen from its backing paper. The paper has a slight coating of a starchy or gluish substance on it that permits the decal to slide off once the paper becomes moist. If the decals are being stubborn, add a few drops of some wetting agent to expedite their release for transfer (e.g., Kodak Photo-Flo-200 solution).

8. Premoisten the area of application and position the decal where it is to be transferred. Carefully slide out the backing material. Press the decal in the center with one finger and gently slide the backing out with the other hand. One may also completely remove the backing prior to application if one finds that a simpler technique.

9. With the aid of a sponge or a small brush, remove all air bubbles between the decal and the ceramic surface. Make certain there is good contact.

10. Let the decal dry overnight. The important thing is to be sure all moisture is removed before firing, or pinholing may develop. If adhesion is at all inadequate, any portions of the decal not in contact with the piece will burn off during the subsequent firing.

11. Fire the piece to Cone 018. Lusters can be applied and fired along with decals. It should take about five hours in the kiln. A slow firing cycle is encouraged in order to avoid blistering, especially in the earlier temperature from 20°C to 426°C (68°F to 800°F). Keep the kiln lid slightly ajar to permit noxious fumes from the medium and the covercoat to escape. Fire only in well-ventilated space.

CONTACT PRINTED PHOTO IMAGES ON CERAMIC SURFACES

Ammonium dichromate (a photo sensitizer) may be mixed with a vehicle or colloid, such as a polyvinyl-alcohol/polyvinyl-acetate solution, gelatin, fish glue, gum arabic, etc. Once combined, this mixture forms a light-sensitive emulsion that can be applied to and processed on a ceramic piece in subdued light. After a piece has been coated and dried, an image can be contact-printed on the emulsion's surface and developed with water, still under subdued light, of course. A negative transparency is needed to reproduce a positive image. One formulary utilizes emulsion ingredients similar to those employed in the gum bichromate process. Instead of adding water-based paint, however, a ceramic pigment (the coloring agent) and glaze base are mixed directly into the emulsion.

The emulsion may also be utilized without any pigment or metallic oxide additive. In this instance, the ammonium or potassium dichromate, when kiln-fired to varying temperatures and under differing conditions, will exhibit a wide range of colors. This is the result of the dichromate's turning to chrome oxide. Thus, not only does ammonium dichromate exhibit photosensitive characteristics, but it also exhibits glaze colorant properties when employed as a metallic oxide ingredient in ceramic decoration.

Contact-printed ceramic images offer an alternative to screened or decal-fabricated images. There are several advantages worth noting:

1. One can print extremely large ceramic pieces. For example, photomurals can be reproduced directly on tiles by utilizing an oversized transparency. It would prove costly, time consuming, and awkward to manipulate a silkscreen to accommodate such a transparency. Extremely large images can be assembled by splicing to-

Untitled. Direct photoceramic image by Kent E. Wade. Kiln-fired on tile using a light-sensitive copper carbonate glaze.

Contact Printed Photoceramic Imagery—Kiln Fired (left to right from top) **1.** *Coat the ceramic surface with the glaze-loaded, sensitized emulsion.* **2.** *Expose the transparency via contact to an ultraviolet light source. Plexiglass is used here to hold the oversized transparency flat. With extra large images, expose the surface in 2 or 3 sections.* **3.** *Develop the image with water aided by a small brush. Exposed tiles are turned over prior to development protecting them from light. After the emulsion is dry, it can be kiln fired.* **4.** *Once the piece is kiln fired, glazed and refired, it can be reassembled. Each tile here is 6" square. Fern Image on Tiles by David Wolaver*

*Right: "Trees" by Kent E. Wade.
Direct photoceramic image,
kiln-fired on tile using a light-
sensitized cobalt glaze. A size to
size transparency was required.
Below: "Flutist" by Kent E. Wade.
Direct photoceramic image,
kiln-fired on tile using a chrome
oxide glaze.*

gether sections of several oversized transparencies. Single-weight glass or a clear plastic sheet will hold the transparency in tight contact with the tiles, and several sun lamps can be employed to make the exposure. One possibility to consider where space is a problem is the assemblage of the tiles outside under a canopy of several sheets of black lightproof plastic. Removal of the canopy will expose the tiles to the ultraviolet rays of the sun. Chromated colloids are "developed" with water. After development, the individual tiles may be fired.

2. More durable, lasting photo images may be produced because one is working mainly with underglaze formulations that are fired in medium-to-high temperature ranges (Cone 06 to Cone 6). Overglaze paints are generally employed for decal fabrication and china painting. They are low-fired (Cone 016 to Cone 018). The interface, or bond, is much stronger between the clay and the glaze in the higher firings; thus, the permanency of the image is greater. The durability of the underglaze image is further increased with the second firing of a transparent glaze over the image. This coat also waterproofs the ceramic surface.

3. Direct ceramic emulsions permit quick and easy application of one-of-a-kind photographic images onto bisque ware (biscuit). The alternatives for photoceramic decoration are further expanded from photoetched plate transfers and photosilkscreened greenware to bisque ware and on to low-fired decal transfers.

The major disadvantage of this process is the difficulty in obtaining a wide variety of repeatable colors. This task would be easier, of course, with the less durable overglaze pigments. Research to date has employed only raw metallic oxides and basic glaze compounds. Spinel oxides and ceramic underglaze stains may offer a greater variety of colors with more controllable results.

There are many variables to be considered in making a ceramic glaze emulsion. Some of the more significant ones are the kiln-firing method employed, the cone temperature of the firing, the amount of the metallic oxide colorant, the type of glaze base, the thickness of the emulsion coating, the surface characteristics of the bisque ware (its texture and degree of vitreousness), the colloid employed, etc. Minor variations in any of these will affect the outcome of the piece. In working out technique, it thus becomes extremely important to maintain accurate records of procedure. This is the only means whereby a duplication of results is possible.

PHOTOCERAMIC TECHNIQUE
Recognizing that this technique is in its early stages of development, the following information is offered as a guide. Pleasing results may be expected.

1. The Emulsion Base. The surface of most bisque ware is porous, although the surface of some unglazed tiles is often quite vitreous. The best colloid for general use, exhibiting optimum adhesive characteristics, is a polyvinyl-alcohol/polyvinyl-acetate solution. Try Screen Star emulsion #71 (regular viscosity), a commercial product that has similar qualities. It is obtainable from silkscreen suppliers. First mix 30 grams of ammonium dichromate with 225 ml of water. This makes a stock solution of sensitizer. Greater quantities of dichromate tend to cause exposure problems with this colloidal base. To make the colloid into a light-sensitive emulsion, mix in ½ part of sensitizer to every 2½ parts of Screen Star emulsion. This emulsion base fires out in the kiln, leaving no noticeable residue.

A series of tests has established that up to a maximum of approximately 50% dry glaze ingredients can be added to this emulsion base without affecting its light-sensitive characteristics. In addition, it has been shown that, given the same quantity of dry glaze ingredients, varying the amount of emulsion base will yield different results. Thus, when determining colors one must be concerned not only with ratio of oxide to glaze base but also with the amount of emulsion base utilized.

When combining ingredients, proceed slowly in order to avoid bubbles, and mix thoroughly, to distribute the glaze particles evenly. Dry raw oxides and glaze ingredients are ground fine enough (about 200 mesh to 400 mesh) as they come packaged at a potters' supply house for addition directly to the emulsion. Add only dry ingredients to the emulsion base; otherwise, excessive bubbles will occur.

On porous bisque ware, one will find that a gum print emulsion will also work to create a photographic stain on pottery. Experimentation will be necessary, and a bit more chromate solution will be helpful. Apply either of these emulsions in subdued light. A negative transparency is used to obtain a positive image.

2. Metallic Oxide Powders and Glaze Bases. The metallic oxides are the chief source of color in the glazes. The color produced will vary with the firing temperature as well as with the fluxes used in the glaze base. Normally, small quantities of an oxide will produce color variation, while increasing these amounts will tend to make the glaze go dark and black. Rather than make up one's own glaze base, try using #101 Clear transparent glaze (dry), sold by the Leslie Ceramic Supply Company. Fire to Cone 04, as the resultant colors achieved at this temperature are more brilliant and exciting. Pieces fired to Cone 4, for example, with an identical application of emulsion, have yielded duller and fewer color variations. It is advisable to run a series of test tiles to establish color ranges with the various oxides and the transparent glaze base selected. Below are listed colors that have been worked out with a certain degree of reliability:

a. Yellow and Browns: Mix 3 grams of red iron oxide plus 3 grams of transparent base #101 plus 15 ml of sensitized emulsion base. Apply in one or two very, very thin coats. Add more emulsion base at any time to any of the mixes, but always be consistent. The less the emulsion, the stronger the image and the heavier the viscosity of the mix with the identical amounts of glaze/colorant materials. A thicker viscosity is a desirable side benefit in controlling the ease with which the emulsion is applied.

b. Blue: Mix 1.5 grams of cobalt carbonate plus 3 grams of transparent base #101 plus 15 ml of sensitized emulsion base. Apply only in a single coat. The image that is left after development gives a reasonably good indication as to what that image will look like after firing. If there are brush strokes or the coating appears uneven, it will fire unevenly. Thus, it becomes very important when applying only a single coating, as is done with cobalt, that the brush strokes be minimized. Have the strokes run in the same direction as the major directional forces within the images themselves. For example, in an image of trees the brush strokes should run vertically with the trunks, not horizontally, counter to them.

c. Blackish-Green: Mix 3 grams of copper carbonate plus 1.5 grams of red iron oxide plus 3 grams of transparent base #101 plus 15 ml of emulsion base. Apply one or two thin coats. When mixing oxides for color variations, the emulsion sometimes becomes too thick to spread evenly and it begins to crack on the surface as it dries. Exposure times may also become excessive. The addition of more emulsion base will help.

d. Green: Mix 1.5 grams of chrome oxide plus 3 grams of transparent base #101 plus 15 ml of emulsion base. Generally, with any of the emulsion blends it is best to apply a thin emulsion coating on a vitreous surface, while a thicker coating is tolerable on more porous surfaces.

e. Clear: A transparent glossy glaze fired over the image will increase its wear resistance. It is best applied in a very thin coat, with separate firing, after the image has been fired on. Choose a glaze that fires to a lower temperature than (not as hot as) that used in affixing the image. The covering glaze can cause the image to bleed, even thin or darken it, depending upon the thickness of the glaze coat, the compatibility of underglaze and overglaze ingredients, and the firing temperature selected. Thus, some knowledge of glazes and testing will be necessary.

3. Special Firing Instructions: As mentioned earlier, it is best to fire in the lower/medium firing ranges in order to obtain richer colors. Each oxide also behaves differently and will create its own intrinsic set of problems at varying temperatures. Do not be too discouraged with the common glaze defects. Remedies are offered in most pottery texts.

Fire tiles in a flat position, rather than upright. There are tile-stacking accessories available for kilns. If tiles are set directly on the kiln shelves, it is best to place some sand under the tiles during firing. This helps to prevent breakage.

The emulsion base gives off a poisonous gas. Therefore, firing should be undertaken in a well-ventilated room. A slow cooling cycle is also recommended.

Lastly, this process utilizes an oxidation firing procedure, although one should not be discouraged from experimenting with a reduction kiln.

4. Coating/Exposure/Development Except for cobalt carbonate (see above, 2.b.), it is generally best to coat the tile or ceramic surface in two layers. In order to achieve an even coat from edge to edge on tiles, place additional tiles on both ends of the one to be coated. Commence application of the emulsion from the middle of one of these tiles, moving across the center tile in one smooth, continuous brush stroke. Apply the emulsion very thinly. The tile will look very streaky. Speed up the drying cycle with a hair dryer set on medium heat. Next, apply a second ultrathin coat, which will be found to cover amazingly well.

Expose the ceramic piece under a negative transparency to any

Making bisque clay slabs for imaging with Direct Photo-emulsion. (left to right from top) ***1.*** *Materials for making non-fired photo images on bisqueware using Rockland Print-E-Mulsion.* ***2.*** *Roll out the clay into an even slab.* ***3.*** *Cut out the shapes desired with a ruler and a metal scribe.* ***4.*** *Bisque fire to Core 05. Photoemulsion can now be applied.*

UV light source and develop the image in water. An identical exposure and development method is further delineated under the gum print process. Exposures may be longer with ceramic applications, however, as the glaze ingredients are very thick.

Generally, one thin coat will hold and fire a halftone image well, while a thicker, double coating is desirable for bolder, more graphic images.

Before coating any ceramic surface, make certain that it is clean and free of grease.

5. Tiles for Photomural Applications: If one is using tiles, double check on the temperature at which the tiles were originally fired in order to eliminate any problems associated with refiring them. In selecting tiles, those with a white background give the strongest contrast for the images.

One may also make one's own tiles or slabs, although it is tricky and warping is a common problem with homemade tiles. The benefits, however, are that clay slabs can be cut up into many irregular shapes. Have the edges of the shapes conform to the stronger lines of the image itself. The pieces can then be reassembled for application of the emulsion, exposure, firing, mounting, etc. Images with a strong graphic design seem to be the most successful with this technique.

The direct photoceramic process is indeed an exciting process, one of special interest for those who enjoy experimentation. There are many colors to be worked out before the full potential of this technique will be realized. In addition, one might also explore the use of ferric salts and diazo compounds as alternatives to dichromated sensitizers (see section on nonsilver processes, Chapter 7).

IMAGE-MAKING WITH CERAMIC POWDERS

This technique is a variation of the dusting-on process (see nonsilver processes, Chapter 7). It is well known that a tacky colloid can be made light-sensitive, so that the light striking the surface of the colloid will make it less tacky. If a positive transparency is inserted between the sensitized surface and the light source, a positive image will be obtained. The image is created by sifting and brushing finely ground powder (such as a ceramic overglaze) over the exposed surface. The powder adheres in direct proportion to the amount of light the surface has received.

The colloid consists of a gelatin-water-sugar mixture sensitized with an ammonium dichromate solution. Use the emulsion in a direct fashion by coating the ceramic surface with a brush or by laying a puddle on the surface of the ware and working it to an even coat. Dry the piece in subdued light. Next, expose the surface to a positive transparency and a UV light source. "Develop" the image with dry ceramic overglaze in powder form. The powder is sprinkled over the surface from an enameling sifter; a small bottle with fine-meshed fabric over its mouth is a good substitute. Lay the powder down in several sprinklings to achieve the desired color intensity. Use a small brush and one's own gentle breath to assist the movement of dust across the surface of the ceramic piece. The image may now be protected with a covercoat. Once the piece is dry, the image can be permanently fired on.

The dusting-on process can also be combined with transfer technique for the application of images to irregular or curved ceramic surfaces. Here, the image is processed on a temporary support, such as glass. After the image has been "developed," it is coated with collodion and the dichromate stain removed. It is then transferred from its support to the ceramic surface much the way a decal is handled. The image is now fused to the piece by firing it at a low temperature in an oxidation kiln.

This process can also be adapted to transfer a powdered-resin resist image to a ceramic or glass substratum. The chosen surface is preheated, permitting the resin to transfer from a temporary paper support. Once the transfer is complete, the image is permanently etched into the surface with acid fumes (review Chapter 2).

Eric Gronborg, "Untitled." Porcelain cups are slab construction with photodecals and lustres applied.

PIGMENTED PHOTORESIST IMAGES ON CERAMIC SURFACES

Kodak Photo Resist (KPR) can be used as a light-sensitive vehicle for the creation of colored images in china paint. As other photoresists may also work, controlled experimentation would be worthwhile.

The selected vitrifiable pigment (i.e., china paint) is milled into a fine powder (220-mesh plus). The pigment is stirred into the KPR in an amount not to exceed an approximate third of the total weight of the combined mix. If there is too much pigment, the resist will not stick to the object.

Before coating the surface, clean it with a degreasing composition, such as a kitchen scouring powder or a household ammonia compound. Next, coat the ceramic piece with the pigmented resist. The resist is exposed in sandwich with a negative transparency to UV light and processed in KPR developer. Exposures will be longer than usual.

After the image is entirely dry, isopropyl (rubbing) alcohol can be used to remove pigment lingering in any unwanted areas. Wrap some cotton on a skewer to assist in this removal process.

It is possible to recoat the surface up to three times using different colors, exposing, developing, and drying between coats.

KPR will fire away, completely and free of ash, at about 426°C (800°F). Follow normal firing methods for overglaze or underglaze adhesion to a ceramic surface. Other fluxes and glaze ingredients may also be mixed with the pigment to increase its fusing characteristics.

Finally, it would appear that this method could easily be adapted to a photoenamel process on metal or glass (review Chapter 1).

Photographic Imagery on Fabric

An exciting way to present a photo image is on textile material. Here the opportunity exists to incorporate an image into articles of clothing, cushions, etc., personalizing the fabric design for additional enjoyment. There are several alternatives for accomplishing this task, ranging from sensitizing solutions and direct photoemulsions, silkscreened dyes and pigmented inks, to photocopier-reproduced image transfers. Let us examine a few of the more interesting techniques and alternatives.

PHOTOSILKSCREENED IMAGES ON FABRIC

This method requires the use of a silkscreen for transferring an image to the textile material. The type of screen and mesh size will vary with the task and the pigment or dye used. Generally a multifilament polyester screen is recommended for textile printing, although a monofilament mesh can be employed. For printing reasonably coarse halftone dots on a smooth fabric, use a 12xx to 16xx mesh. Rougher fabrics will require about a 10xx or comparable mesh size. In textile printing, inks tend to spread, so one should avoid laying down an excessive quantity of ink. Because of the various combinations of mesh fabrics, inks, and textile materials available, one should consult with a dye and fabric dealer or silkscreen supplier for specific recommendations when utilizing their materials. For textile printing, use a direct photoemulsion: Naz-Dar Encosol-3 for water-based inks and Encosol-1 or Encosol-2 for oil-based products. To prepare the mesh for printing, follow basic photosilkscreen technique (see Chapter 9).

There is a wide variety of dyes and pigments that can be employed for printing on textiles. Dyes add color to the fabric by chemically altering the characteristics of the fiber itself. Pigment-based inks, on the other hand, are dependent upon binder additives to help encompass the fiber strands in a sheath of color. Dyes hold up well in washing and are very economical. Dyes also offer brilliant

Photosilkscreening a piece of fabric: (left to right from top)
1. Applying the registration paste to stiff cardboard. 2. Rolling out the fabric on the tacky board.
3. The photosilkscreened image is registered. 4. The image is squeegeed onto the fabric. 5. The image is transferred to the fabric.

colors, as do pigmented inks; however, some pigments tend to afford better shades of green and blue. Pigments are considered more lightfast and can be purchased in a prepared state, ready for screening. Several interesting dyes and pigment-based inks that are suitable for photosilkscreening are listed below:

1. Fiber-Reactive Dyes. These dyes are packaged in a dry powder form under several brand names (e.g., Procion). They are easy to use, but they must first be mixed up into a base solution, the proportion of ingredients determined by the use and type of dye. These solutions usually contain some: **a.** urea, a hydroscopic agent to prevent image-bleeding; **b.** the sodium salt of M-nitrobenzenesulfonic acid, an oxidizing agent, to maximize the intensity of the color; **c.** distilled water or softened tap water (e.g., Calgon-treated) to neutralize unwanted metallic salts, and **d.** sodium alginate, a seaweed derivative used as a thickener for silkscreening purposes. These chemicals, or premeasured thickener/base kits, are available from textile-printing suppliers. These highly concentrated dyes come in a wide range of colors and can be used only on natural fiber materials, such as 100% untreated cotton, silk, leather, wood, or paper. A rich color is attained with dye quantities representing 5% to 10% of the screening mix. For general safety, one should wear gloves when handling dye materials and avoid breathing dye powders.

2. Disperse Dyes. These dyes are packaged in either a liquid (predispersed) or a dry powder form under several labels (e.g., Dispersol Dyes). They come in several brilliant colors and may be employed for photosilkscreening on nylon, acetate, and polyester fabrics. These dyes are made colorfast to light and water by processing with heat and pressure at low temperatures (135°C; 275°F). The dyes sublime; that is, they are modified from a dry to a gaseous state during processing. The fibers open up with heat and encapsulate the fumes given off by the dye. The gas is thus entrapped by the fiber, not absorbed by it.

Disperse dyes, because of their sublimation qualities, are also well suited to transfer printing techniques, in addition to the photo-silkscreening of images. The 3M Company markets the Color-in-Color copier machine (to be discontinued, yet widely available), which will reproduce full-color imagery on matrix paper, making possible image transfers to fabric and other surfaces. To transfer images to natural fibrous materials (100% cotton, wood, paper, etc.), the surface must first be sprayed or coated with a polyvinyl chloride chemical or a polyester-type solution that promotes transfer of the dyes. Metal surfaces can have color images transferred to them if they are first coated evenly with a solution based on methyl ethyl ketone (MEK). Some ceramic companies use a polyester release coating in their production molds. Ceramic pieces thus manufactured will also accept the dyes. Plexiglas, polyester fabrics, and pretreated acetate surfaces do not require this special treatment. The image is transferred directly from the matrix paper via heat application (137°C; 280°F).

3M Color Layout Sheets, precoated with disperse dyes, have been used successfully to transfer sublimation dye images to fabric utilizing an IBM photocopier. The photograph is placed on the copier's platen and a reasonable facsimile of the original is reproduced as a black image on a white paper base. The dye sheet and the copy are then sandwiched together (image to dye side) and sublimed in a dry-mount press for about 12 seconds. The dye sheet is then peeled away. The black image areas of the copy attract the dye. The nonimage areas also attract the dye, yet not to the deep intensity of the black image areas. The photocopy is next brought into contact with a piece of fabric, such as a tightly woven polyester or a multi-filament Dacron, and again sublimed in a dry-mount press. This completes the transfer of the image. Prior to transfer of the image to the fabric, tinted backgrounds may be cut away, or several images combined. Although this is basically a monochromatic technique, additional colors might be added to any one image via re-registration methods and heat-application technique.

It is feasible to make one's own dye sheets. Experiment with Patina Coated Matte or Z-Heat Transfer paper (Zellerbach Paper Co.), which are special papers on which one can screen subliming dyes for heat-transfer to synthetic fabric. Other intermediate surfaces should also be explored, such as clay-coated printing stock, label, or any ordinary 50# to 60# smooth-surfaced paper, etc. Once a suitable release paper is chosen, it could be open-screened or brushed with dye for heat-transfer of photocopy images to the fabric. One could also use PCM paper (Patina Coated Matte) to photosilkscreen or produce lithographically a four-color image, using transparent disperse dyes for eventual transfer to synthetic fabric. When an image is being silkscreened onto a release paper, a 200-mesh monofilament polyester can be employed. Images thus produced are sharper than those printed with a multi-filament mesh directly on the fabric. Try

The top photo shows a color layout or dye sheet being peeled away from an IBM photocopy after having been subjected to heat in a drymount press for 10 seconds. Bottom photo: The blackened image areas of the photostatic copy attracted the dye which is transferred again via heat to fabric (top sheet). Photos courtesy of Diane Sheehan.

Transcello process colors (Hit series) or Colonial Trans-Fab Inks for silkscreening on synthetics. When transferring an image to natural fibers, try coating the image first with a clear Plasticol ink to promote dye acceptance by the fibers.

There is a product, manufactured by Binney & Smith, Inc., of New York, called Crayola Craft Fabric Crayon, No. 5008, which contains sublimation dyes in a wax base. Photographic images can be quickly reproduced on a black-and-white photocopier and colored by hand with any one of eight colors. Once the image is completed, place the colored image in contact with a polyester fabric. Any household iron can be used to heat-transfer the image. Only the dye will transfer to the material. These crayons might also be utilized to hand-color images produced by other means for eventual transfer to fabric.

3. Water-Based Pigmented Inks. There are several water-based pigmented inks available (e.g., Watertex, Versatex, etc.). These are designed primarily for silkscreening. They are packaged in a variety of transparent colors, the translucence of which may be altered with the appropriate extender. It is important to remember, when applying any transparent ink, that the color underneath will be visible, thus creating a third color. If the fabric is already colored, then consideration should be given to screening one area of the fabric white before laying down the colors for the image. Binder agents are also available. They are mixed directly with the ink and increase the durability of the image by improving the holding power of the color. Once the fabric has been screened, heat should be employed to set the image permanently. This can be done by ironing the reverse side of the fabric. These inks can be used on natural fiber as well as on some synthetic materials without stiffening the fabric. When preparing the silkscreen, be sure to use a waterproof photoemulsion.

4. Light-Sensitive Inkodye. Inkodye colors are reasonably perma-

Light sensitive Inko Dye materials.

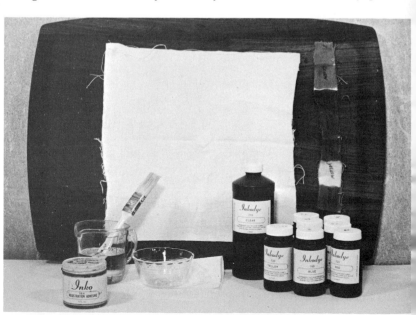

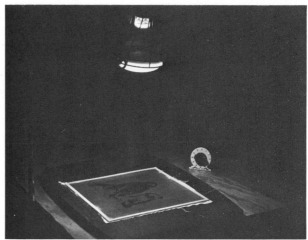

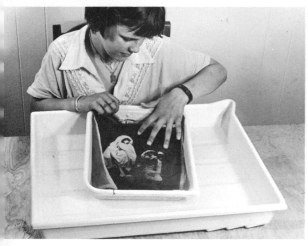

nent vat dyes in a leuco base form. They do tend to fade slightly, however, if exposed to direct sunlight for extended periods of time. They are supplied in a variety of transparent colors and may be intermixed. An extender is also available to lighten values, although water can be substituted. These dyes can be used on untreated natural fibers, including paper and leather. The dye must be handled under subdued light and stored in opaque bottles.

In order to photosilkscreen the dye, a water-resistant direct emulsion must be employed. In addition, a multifilament polyester, mesh 10xx to 14xx, is recommended. As one screens the image, it may be necessary to pull the squeegee two or three times to insure good dye penetration and coverage. One may print overlapping multiple colors, as long as the underneath color has completely dried first. This technique does produce secondary colors and should be given some consideration prior to actual screening of the dye. The image is set and developed all at one time by exposure to sunlight or to a hot iron. Sunlight gives more brilliant colors. Once the dyes are developed, merely wash the fabric in warm soapy water,

Procedures for imaging with Inko Dyes:(left to right from top)
1. Secure the fabric to a support. Coat in subdued light and let dry. The dye does not obtain its color until after exposure to light.
2. Place negative transparency over the sensitized fabric and sandwich together with a sheet of glass. Expose to a sun lamp or other ultraviolet light source.
3. Remove the unexposed areas with a water development. Change water several times until all traces of the unexposed dye are removed. Reapply another dye color, if desired. 4. Set the image by ironing until no more fumes are given off.

The final image is viewed here by transmitted light to accentuate the material. This is the effect obtained by hanging a cloth image in a window.

rinse, and lay flat to dry. For silk fabrics, one should use Inko Silk Dyes.

The light-sensitive characteristics of Inkodyes may be exploited photographically in order to produce images directly, by-passing the silkscreen operation. Brush the dye on the fabric under low light. As soon as the fabric has dried, smooth it out and place a negative transparency on the surface (emulsion side to dye side). Next, place a sheet of glass over the top and expose to any ultraviolet light source. Each color seems to have its own speed at which the light brings it to full color saturation. For example, with a sun lamp at approximately 65cm (25 in), the undiluted yellow and violet colors darken within 3 minutes, the red and blue within 15 minutes, and the green and brown within 25-to-30 minutes. If the fabric has wet spots prior to exposure, those areas will elicit uneven tones. (Ultimately, a test strip should be run with any dye dilution, in order to determine an optimum exposure range.) To develop the image, immerse the fabric in cold water until no more coloring washes out. Next, dry the fabric in a horizontal position to prevent any residual bleeding. Once the material is dry, apply a hot iron to the reverse side of the material, (not the treated surface, in order to avoid scorching of the image). This helps to set the remaining dye. The colors will tend to deepen slightly during this process. Keep ironing until the dye stops fuming. Fumes are nontoxic, but they can be disagreeable; therefore, iron in a well-ventilated area.

FABRIC SELECTION AND REGISTRATION TECHNIQUE
The fiber content of the material selected should first be ascertained so that its compatability with various dyes or pigments can be established. It may be necessary to alter one's choice of fabric or dye should the two not prove compatible.

For inks necessitating a nonsynthetic material, it is best to use only bleached-white, 100% natural fibers. Make certain that any sizing has been removed. (The sizing will affect penetration of the ink.) If necessary, one may remove sizing with the aid of a detergent wash. Generally it is best to select fabric that is unsized and untreated, although some cotton materials that have been mercerized are worthy of attention. Fabric that has been mercerized will yield richer, more vivid colors, as the fibers tend to display a greater attraction for the dye molecules. Fabric may be obtained from most dye suppliers. Before actually screening the material, preshrink it by washing in hot water. This will help avoid any future registration problems.

It is advisable that the fabric be laid out smoothly under slight tension prior to photosilkscreening the dye or pigment, in order to minimize wrinkles. One method that will improve the accuracy of registration of several colors, a method that is suited to multiple-color printing, uses an adhesive paste. Try Inko #184 Registration Adhesive. A little bit goes a long way. Dilute some paste with turpentine and spread a thin, yet even, coating onto a piece of cardboard. After the surface has dried for about 15 minutes, roll out the fabric onto the tacky surface and smooth it with the hand. The tension will now be distributed more evenly over the entire fabric, and the fabric will be less apt to bunch up. Sharper images will be obtained and re-registration of the image for each color will be more easily achieved. Once printing is completed, the fabric can easily be removed from the cardboard.

PHOTOFABRIC EMULSIONS

Sometimes fabric may be used as a support for a photographic emulsion. Because these emulsions contain a colloid base (e.g., gelatin), a fabric that has been coated with an emulsion tends to be rather stiff. This somewhat limits the use of the material for items such as wearing apparel. There are still, however, many other textile uses (such as soft sculptural pieces) for which fabric rigidity might not necessarily be a problem.

What are the available options for photoemulsion images on fabric? One is to purchase a photographic linen (e.g., Luminos Photo Linen). The base stock is a linen-canvas material, available in sheets or in rolls. The material can be washed, sewn, twisted, hand-colored, etc. The emulsion is a bromide type that processes like a conventional black-and-white photographic paper. The blueprint industry also markets a light-sensitive fabric for architects, although it is less readily available because of the rising cost of linen.

One may purchase liquid photographic emulsions for application to cloth. Try Rockland's Fast Enlargement Speed Emulsion, BB-201, on canvas. This product can be used for making wall tapestries photographically, as well as for general projection printing on fabric. When using the emulsion, maintain as even a thickness as possible. A spray gun may offer better control during application of the emulsion, especially on larger surfaces. The emulsion processes the same way photographic paper does, although the exposures are significantly longer (review photoceramics chapter for processing technique). If one prefers, it is possible to return to basics and make a homemade emulsion, although this can be very time-consuming compared to available alternatives.

PHOTOSENSITIZERS FOR FABRIC

Inkodyes were mentioned earlier as one type of photosensitizer. There are many other methods and formulas for photosensitizing cloth. Let's examine a few of them.

a. Ferric-Silver Sensitizers. Ferric-silver solutions such as the Vandyke brownprint sensitizer enable the user to saturate a fabric with a sensitizing dye without obtaining the stiffness typically associated with photoemulsions. These compounds should be used principally with unsized nonsynthetic fabrics or paper. When a negative transparency is placed in contact with the sensitized material and exposed to an ultraviolet light source (sun, unfiltered black light, etc.), the ferric salts will combine with the silver nitrate, precipitating out the metallic silver to form the photo image. These formulas offer a limited color range (between brown and black), depending upon developing treatment.

To make a brownprint sensitizer, you will need:
- 1. 84.6 grams ferric ammonium citrate (green)
- 2. 14 grams tartaric acid
- 3. 35 grams silver nitrate*
- 4. 1 liter distilled water

***WARNING:** Wear rubber gloves when working with silver nitrate, as it can cause burns to skin and eyes. It also will leave stubborn, dark stains on skin.

Dissolve each chemical separately into approximately 300 ml of the distilled water. Next, mix together items 1 and 2. Then slowly stir in item 3, using a glass rod. Add the remaining 100 ml of the distilled water. Pour this solution into an amber-colored glass bottle for storage in a cool, dark place. Under the proper conditions, this sensitizer will last for two or three months.

The fabric is best coated either by soaking it for about four minutes in the sensitizing solution or by applying the solution heavily with a soft brush or a Blanchard Brush (which is merely a piece of soft cotton flannel wrapped several times about a strip of glass, comfortable to hold in the hand). When brushing the fabric, apply the sensitizer with both horizontal and vertical strokes. All coating should be done under a yellow or other subdued light source (such as a yellow insect-repellent light). Hang the fabric to dry in the dark, turning it end for end once or twice so that it dries evenly.

Once the fabric is completely dry, place it on top of a black surface, sensitized side up. This is a negative-positive process, therefore position a negative transparency on top of the fabric and cover it with a sheet of glass to form a sandwich. It is advisable to use a normal or a contrasty transparency, as low-contrast images tend to give unsatisfactory results.

Expose to any ultraviolet light source (e.g., sun lamp at 65cm, or 25 inches, approximately 20 minutes). A standard photographic test strip is always useful to ascertain optimum exposures. A useful exposure is normally obtained when the highlight areas begin to generate detail. The longer the exposure, the darker the browns will be.

Once the image has been exposed, wash the print in running water at room temperature for approximately one minute. The fabric should next be fixed in a hypo bath (28 grams of sodium thiosulfate to 600ml of water) where it will turn from a yellow to a Vandyke brown. Be careful not to overfix, as images tend to reduce slightly in the hypo. Once the desired tone is achieved, wash the print again for 15 to 30 minutes in running water and hang the fabric to dry. The print may also be fixed by drying the fabric in direct sunlight after the initial rinse. Another brown tone will be obtained. The print color may be further altered from brown to black by ironing the image while the fabric is still damp from the wash.

A variation to the brownprint sensitizer can be made as follows:
- 1. 2.6 grams of ferric ammonium citrate (green)
- 2. 0.4 grams of oxalic acid*
- 3. 17 grains of silver nitrate**
- 4. 35ml distilled water

*CAUTION: It can cause skin irritations.
**Beware of burns and stains to the skin.

Mix items 1, 2, and 4 together, using a glass rod, and slowly add in item 3. Use this solution to coat the fabric under yellow safelight. Once the cloth is dry, position the transparency and expose to any UV light source. The emulsion will turn from yellow to a brownish tint.

Develop the fabric for approximately five minutes in a borax solution (a tablespoon of borax to a liter of tap water). Next, the image is fixed for approximately six minutes in one tablespoon of sodium thiosulfate dissolved in a liter of tap water. The fabric may now be washed for 20 minutes and hung to dry.

This formula can also be modified for use in sensitizing paper, as follows:

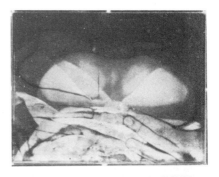

Take stock solution A (composed of items 1, 2, and 4 above) and combine with a weak gelatin-and-water mixture, stock solution B. Stock solution B is made by mixing 11 grains of gelatin into 10ml water. Just prior to use, dilute stock solution B with enough hot water to give it a fluid consistency and mix in stock solution A. Then slowly add in the silver nitrate (item 3 above) and mix the ingredients together thoroughly. Spread immediately on the paper under safelight conditions, expose, and process as described for fabric.

The Rockland Colloid Company markets a product called Rockland Fabric Sensitizer FA-1, which yields results similar to the above ferric-silver sensitizers. It is also a contact-speed sensitizer and should be used only on 100% natural fiber. The kit contains all the chemicals necessary to make one gallon of working solution. There are two parts to the sensitizer, 1a and 1b. Dissolve 1a and 1b separately in 1.9 liters (½ gallon) of distilled water and store in amber bottles. Prior to use, mix equal parts of 1a and 1b under subdued light. Next, soak the fabric in the sensitizer, dry it, position the transparency and expose it to any UV light source. Develop the exposed fabric in water for approximately one minute, then place it in the hypo solution (part 2 of the kit) for approximately 20 seconds.

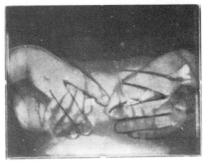

The fabric should next be washed for 20 to 30 minutes to remove any unprocessed chemical. The image thus obtained is a reasonably permanent gray-brown color. The fabric can be washed; however, strong detergents are to be avoided.

b. Ferric-Salts Sensitizers. The cyanotype formula detailed under Nonsilver Processes may also be utilized on 100% natural fiber material, preferably bleached, to produce a blue image on a white ground. A negative transparency is needed to produce a positive image. The fabric is best soaked in the sensitizer for approximately four minutes to obtain an even coat and maximum saturation. Once the fabric has dried, it may be exposed to a UV light source, then developed and fixed in water. The image produced is reasonably permanent, and the fabric will not be stiffened by the salts.

c. Diazotype Textile Printing and Mordant Dye Printing. In diazotype printing, a positive transparency is utilized to obtain a positive image. Some choice of color is possible, depending on the developer employed. The basic steps consist of sizing the fabric, sensitizing it, exposing it to a UV light source while still wet, and then soaking it in the appropriate developer to obtain the desired color: red, yellow, purple, etc. Unfortunately, colors produced in this manner are considered low in saturation.

Another similar type of process (phototincture) is the mordant dye printing technique. Here, the fabric is treated with an ammonium dichromate solution (light-sensitive metallic salt) and then exposed to ultraviolet light. A negative transparency is placed in contact with the fabric during exposure. The light causes the formation of chromium hydroxide, which acts as a mordant to hold fast certain dyes, such as alizarine and anthracine. There are other natural dyestuffs that utilize potassium dichromate as the mordant. These are certainly worth exploring and may be obtained from dye and fabric suppliers.

XEROX COLOR COPIER PHOTOGRAPHIC TRANSFERS

Information on the Xerox color copier machine is presented in a later chapter; however, it is important to note that the Xerox 6500 Color Copier may be used to reproduce color photos and slide projections, as well as to make two-dimensional representations of three-dimensional objects, up to 22x36cm (9x14 in), suitable for transfer to fabric. One will need a thermoplastic transfer paper to receive the colored toners used in the copier. Zellerbach's Transeal Heat Transfer Paper 300-J can be used successfully for this purpose with perhaps minor adjustments to the copier's paper-feed mechanism. The image is reproduced directly on the release paper, where it is temporarily fused to it at approximately 210°C (410°F). When the image is brought into contact with a 100% cotton fabric in a dry-mount press at temperatures slightly above the original fusing point of the toner, the image will transfer to the fabric. Remember to reverse the color slide before making the copy, so that it will read correctly after the transfer. This system is very quick and affords the user a great deal of flexibility in reproducing imagery more difficult to attain with other methods.

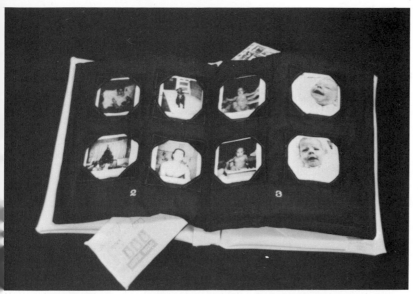

Above left: Catherine Jansen, "Family Album." Images were made using the 3M Color-in-Color machine. The soft album has been stuffed and stitched to give the piece a 3-dimensional quality. Below left: Judy Lehner "Grandma." Brownprint (left) was created using Rockland FA—a fabric sensitizer printed on 100% cotton. The transparency for the brownprint was made from the original print (right) which was hand colored in oils about 1870. The brownprint has been hand-painted using Procoin dyes and further embellished with pink french knots. It is framed in a pink/beige binding and lace.

PRESENTATION OF PHOTO IMAGES ON TEXTILES

Several methods for printing on cloth have been discussed. Basically, all these techniques appear to accomplish the same end. Some methods offer more stable images, while some offer more colorful images. Once the image is integrated with the fabric, there are several things that may be done to enhance presentation of the image. An image on cloth can be colored by hand with a variety of different substances, overlaid with dyed photographic transparencies, etc., in order to lend additional vibrancy to the fabric and to the image. By stitching with a contrasting colored thread around several shapes in an image, a new dimension can be added to it. This accenting of the image can be further reinforced by stuffing the fabric either selectively, under certain portions of the image, or completely, under the entire image. If one intends to hang the work, one might include such ideas as canvas stretcher bars, thick plastic dowel rods at top and bottom (to create tension on the fabric), or perhaps an oversized embroidery hoop to stretch and hold the cloth. The piece could then be hung in a window like a stained-glass design. There are many possibilities for the final photo-fabricated-image. One might create stuffed animals, furniture—and even friends!

William Larson, "Class Piece." The piece consists of over 200 embroidered words and other needlework on fabric. The images were created using the blueprint technique, then stuffed, stitched, and zippered together. 25" x 14". (Courtesy, Light Gallery)

Detail "Elfia Muenzer"

Detail "Martin Buddelman"

Printing and Etching of Other Nontraditional Surfaces

Some of the techniques mentioned up to this point can be used for printing on surfaces other than metal, glass, ceramics, and cloth. Some of the other more common surfaces, and interesting methods of photographic application of imagery, are delineated below.

WOOD AND LINOLEUM

Any liquid photoemulsion can be used for creating images on wood. Before applying the emulsion to hardwood, it is advisable to apply a thin coat of polyurethane to the surface. This increases surface tension and bonding ability of the emulsion. It may also prove advantageous to paint a white background on the surface of the wood. The reason for this is that the image develops out black and contrasts with the natural color of the wood. If the wood is too dark, then the image may be lost. In addition to using a photographic emulsion, photosilkscreen technique can be used to render an image successfully on wood. Here one has an assortment of pigments and a virtually unlimited choice of colors.

If one is interested in creating a photographic woodcut for block printing and/or display, the photoemulsion can be very useful in creating the pattern. A linoleum block can also be used. The image, once photographically reproduced on the block, can be chiseled out with a cutter or carved with a wood-burning iron, "developing" the image in either relief or intaglio for printing. Printing technique can take the form of more traditional block-print inking methods. Once the cutting or burning operation is complete, brush off any loose bits of wood. Next, use a soft rubber brayer to roll a coat of block-printing ink evenly over the image. Place a sheet of paper over the inked relief and rub the back of the paper with a wooden spoon. Pressure variations will create light and dark tones in the image. This effect can be controlled and used to advantage. Once the image has been burnished, the print can be removed and mounted.

A homemade silver sensitizer can also be used to create the

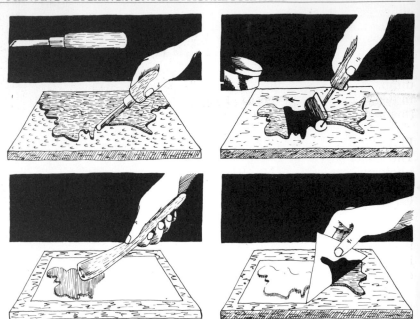

Making a photo woodcut transfer: (left to right from top) **1.** *The photoimage is burned into the wood (intaglio).* **2.** *The photoimage is carved in relief on the wood.* **3.** *The carved out photoimage is inked in traditional woodcut fashion.* **4.** *The image is transferred to the paper by burnishing the back of the paper.* **5.** *The image is transferred and the paper removed. The woodcut may be inked and reused.*

image/pattern on wood. Prepare a base coat for the sensitizer by whipping several egg whites (albumin) into a foam. Once the albumin has settled to a liquid state, add zinc oxide until a thin, smooth consistency is obtained. Brush a thin coating of this solution onto a block of wood. After it dries, pour the sensitizer over the surface in subdued light. The sensitizer is made by mixing approximately 1.5 grams of silver nitrate in 16ml of water. Let the surface dry again. A transparency can now be sandwiched between the wooden block and a piece of glass and exposed to any UV light source. Fix the image in a weak sodium thiosulfate solution (30 grams of sodium thiosulfate mixed with 600ml of water). Next, wash and dry the surface. The image can then be chiseled out in relief. The brownprint sensitizer delineated in the last chapter (for images on fabric) affords yet another possibility.

A photographic image may be transferred to a piece of wood by sandblasting. Coat the wood in several layers with an undiluted direct photosilkscreen emulsion (review photo-sandblast-resists for glass). This should be undertaken in subdued light. Place a high-contrast transparency on the emulsion and sandwich it with a piece of glass. The emulsion is exposed and then processed in water. Next do any necessary touchup to the resist coating, dry it, and sandblast the image. A very thick coating is desirable, as the resist will eventually yield to the abrasive particles. Softwood (e.g., pine) blasts very easily, but the broad grain pattern is visually distracting. Preferably, one should use a more tightly grained hardwood to eliminate this problem. The image may be created either in intaglio or in relief. A relief image could be employed as a printing block for making woodcut prints.

It would seem feasible that a lino cut could also be produced photographically using a photoresist technique, processing with either sand or acid. In the latter instance, an intermediate resist barrier may be necessary. In addition, a dilute sodium hydroxide (caustic soda) solution would be employed as the etch. (*WARNING:* This compound is dangerous and must be handled carefully.) Asphaltum could be used as a suitable block-out and touch-up medium, should this be necessary. A lino cut can be inked and printed in the woodcut technique. It is also feasible to use an etching press (medium pressure) to transfer the image to paper or leather.

PLASTICS

The major process to consider when working on plastics is the photosilkscreening technique. There are several excellent inks for plastics (e.g., Naz-Dar) that exhibit excellent bonding characteristics yet are well suited to the rigors of further manipulation. Plastic can be easily cut or shaped. It can also be molded by the application of heat (vacuum forming technique). If one decides to utilize an image in a creative way that requires reshaping of the plastic sheet, then the inks must be able to withstand these stresses. It is always best to consult a silkscreen dealer and to choose the ink to meet the task.

The transparent quality of plastic may also be exploited effectively as regards image presentation. Using very thin sheets of acetate or Mylar, posterized components of an image may be screened on individual sheets, then overlaid upon or behind each other. This may be done with several colored inks, creating effects similar to those obtained with color-proofing films.

If one is seeking a continuous-tone black-and-white reproduction without employing a halftone or texture screen, then a liquid photoemulsion may be the answer. As previously stated, this product will coat almost any surface. Its major drawbacks are, of course, its susceptibility to marring and chipping. Finally, a photosensitive sandblast resist, such as that used on glass surfaces, may be adapted for use on plastic. Here, the abrasive particles create a translucent impression by altering the transparency of the plastic.

RUBBER

Photographic images can be reproduced in rubber, either for display or for use as a transfer tool (a stamp). This technique first entails photoetching a metal plate by standard practices. Etch the plate to a depth of approximately 3mm (1/8 in). Depending on the effect one hopes to achieve, use either a positive or a negative transparency to create the image. Line images will work better than halftones, although halftone images as fine as those attained using an 80-line screen can be reproduced in rubber. If halftone images are to be used as rubber stamps, experiment with a coarser dot pattern (40-line to 60-line) until stamping technique is perfected.

Once the photoetching is prepared, the following procedure may be used to make a vulcanized rubber photographic relief:

Preheat the photoetched plate in a rubber-stamp press for two-to-five minutes at 145°C to 162°C (300°F - 325°F) under zero pressure. Next, cut a piece of matrix material from a sheet of Bakelite (Consolidated Stamp Co.) the size of the plate. Place this matrix material on top of the etched side of the plate and insert both in the press. A sheet

Top photo: Jackie Leventhal, rubber stamp (left side) is hand crafted from center B & W photograph.

photograph. Postage stamp (right) made from same photo.

Bottom: Jackie Leventhal. At left is a rubber stamp, a photoetched metal plate fastened to a wood block, and a rubber stamp print. Center top is the original B & W photograph. Center bottom depicts 3 rubber stamp print variations utilizing several different color inks. Center middle is a Bakelite matrix made from the metal plate. Rubber gum is placed in contact with the Bakelite to obtain the rubber stamp. Right side, top is an enlarged Kodalith transparency of the B & W photograph. Right side bottom is a B & W print from the transparency. Note that the transparency is not the size used to make the rubber stamp. A reduced, high contrast transparency was employed to photoetch the metal plate. The rubber stamp is 4" x 6".

of holland cloth (a thin, tightly woven fabric) is placed over the matrix to protect the press. Close the press for 1½ minutes with zero pressure applied. Next, increase the pressure to approximately 1500 pounds for five minutes. Heat and uniform pressure are required to create a satisfactory mold.

The image is now transferred from the matrix to Gum Stamp Rubber or even recycled tire rubber (3mm to 5mm—1/8 in to 3/16 in—thick). Cut a piece identical in size to the matrix. If Gum Stamp Rubber is used, peel off the backing and affix the slightly tacky side to the image side of the matrix. Place a piece of holland cloth over the rubber to protect the top of the press and to subject the matrix/rubber sandwich to heat and pressure for approximately three-to-five minutes. The matrix can be used more than once.

The rubber-based relief image may now be cut out and mounted. If one intends to make a stamp, it is advisable to cushion the back of the piece with a thin slice of medium soft rubber before mounting. A handle device can be designed out of ceramic, hardwood, etc.

There is a variety of substances for inking a multitude of surfaces. For example, try one of the rubber-stamp compounds designed for etching glass (Etching Ink No. 2, L. Reusch & Co.) or some of the colorful inks designed for paper. One is encouraged also to explore the use of homemade formulations, as well as masking or using selective inking technique in order to create multicolored imagery. With halftone images, use a stiffer inking pad to minimize a potentially heavy ink release that could result in a splotchy impression.

This process is a relatively simple one. Its biggest drawback is that it requires a rubber-stamp press, which is reasonable in cost for what it is, yet expensive. It would seem feasible, however, that an economical heat-and-pressure unit could be made with a little ingenuity.

STONE

A photoresist may be used on slate to create photographic images in stone. The etchant employed is hydrofluoric acid, and therefore one must be *extremely CAUTIOUS* (review etching glass). Prior to sensitizing the slate, try ball-milling Kodak KTFR as a substitute resist in the Kodak Metal-Etch Resist Additive D Formula. Once the solution has nearly cleared, it may be applied to the stone under subdued light, exposed, processed, and the slate etched.

Various solutions using hydrochloric acid will etch marble. A modified resist could be utilized on this, as well as on similar stone surfaces. If it suits one's need, stone can also be coated with a commercially made gelatin-based chlorobromide emulsion. There is the possibility, too, of sandblasting soft rock such as soapstone. A photosensitive sandblast resist for use with glass could easily be adapted for such surfaces.

SKIN IMAGES: FOTO FICTION

Photoimages on skin surfaces have traditionally been produced on animal hides via an etching press or various photographic sensitizers. These processes do not lend themselves to human application; therefore, a contemporary form of skin printing is being presented here as a legitimate alternative for the experimental photographer.

The following is a contact process utilizing a high-contrast transparency, although wrist watches, rings, hands, etc., may be employed. When selecting material, be cautious of printing areas that contain excessive freckles. These small dots tend to break up the image, giving it a very grainy effect. Fair-skinned bodies that burn easily in direct sunlight work best. To prepare the printing surface, have the subject remain indoors for several weeks, bathing continually. This should prove sufficient to render the skin tone a bleached white. If one is unable to secure any body willing to be subjected to the rigors of photo preparation, one may have to settle for more delicate posterior images. Here the opportunity for creating double exposures is evident.

A photoetched image on slate using Kodak Kmer Additive D formula plus Hydrofluoric acid. By Charles A. Bigelow.

Once the body is conditioned, position the transparency and expose the subject to direct sunlight. Those people living in the Northwest may have to use artificial UV light sources. After the exposure is complete, develop the image in baby oil. Development serves several purposes: it tends to darken the image, it prevents premature image deterioration due to flaking, and it helps to reduce any pain associated with overexposure. The image may now be rinsed in running water for five minutes, air dried, and hand-colored with a variety of substances.

As an alternative to the above, some consideration should be given to conventional photosilkscreening using water-based inks—as well as to this writer's attempt at foto fiction.

Beyond the Black and White

There are many visually interesting ways to render a silver image effectively. Once again, it is found that the techniques are not necessarily new, only a fresh approach to the employment of the tools themselves is being recognized. Some of these methods would by contemporary standards definitely be considered unconventional; nonpurist, to say the least. Regardless of their status, many will be found artistically rewarding. Five areas will be examined: bleaching of silver bromide paper, addition of color to black-and-white prints, the dye imbibition process, litho-etch/dye-staining technique, and various transfer techniques.

BLEACHING THE SILVER PRINT
Bleaching is a removal process. The creative use of reduction compounds can add an entire new dimension to one's images. All that is needed is a silver bromide print and a bleaching solution. Today, bleaches such as Farmer's Reducer are mainly employed to dissolve the silver salts in the print, thus rendering whites "whiter than white" as well as lightening up dense shadow areas. Our interest, however, is in the controlled usage of bleach and of some of its traditionally undesirable side effects. Some of these possibilities are as follows:

1. Farmer's Reducer tends to leave a brownish-yellowish stain. This stain may be incorporated into an image or employed as a tinting agent.

2. Sometimes a photograph can be given a surreal effect by drawing or sketching over segments of the image with a pencil. The graphite acts as a resist. Detail may even be added while bleaching away the silver image. All that remains is the drawing or a combined drawing/black-and-white print, depending on the amount of bleaching. Smudging the graphite also gives an interesting variation.

3. Bleach can be used selectively by first masking out certain portions of the image with rubber cement or with a frisket, such as

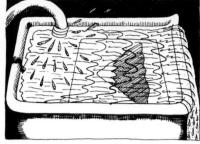

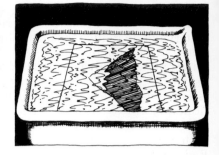

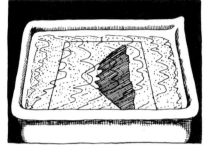

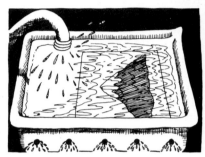

Bleaching a print: (left to right from top).
1. *Part A (Kodak Fixer) plus part B (Potassium Ferricyanide) = bleach bath. Mix together just prior to use. Process for 2 minutes.* ***2.*** *Rinse after bleach bath, 45 seconds.*
3. *Fixer, 6 minutes.* ***4.*** *Hypo Eliminator bath, 2 minutes.* ***5.*** *Wash in running water, 5 minutes.*

Grumbacher's Miskit, and then bleaching the print. The effects can be startling. Bleach also can be applied effectively with a small brush, one without a metal ferrule because bleach reacts aggressively on contact with metal. Try using a Chinese bamboo watercolor brush.

4. Overexposing a print quite heavily (especially one with a dark background and a light subject) and then bleaching it almost to paper base will tend to create an extra color—brown. Experiment with bleaching solarized prints. Grainy images work very well. With bleaching techniques, one can create some beautiful imagery out of black-and-white rejects.

5. Try bleaching out colored, or monochromatic, photographic paper. Also try multiple-toning prints in conjunction with bleaching technique.

Bleaching Solutions

Three basic bleaching formulas are presented here, although there are many different solutions available. These three are mentioned because they have been used successfully in the reduction of prints.

1. Farmer's Reducer. Kodak markets small packets of Farmer's Reducer that are available at most photo stores. It is just as easy, however, to make up a brew from scratch. Two solutions are required. A: Print fixer. Mix this up as one would for normal darkroom usage. The main function of the fixer, or hypo, is to remove the silver ferricyanide, a by-product of the potassium ferricyanide's reducing action. It does permit observation of the reducing action in a single-bath system and it does neutralize the action of the bleach in a short period of time when mixed together with part B. B: Potassium ferricyanide, the bleach, diluted with water. The concentration of potassium ferricyanide will determine the cutting action of the solution. Weak dilutions, such as 7 grams of potassium ferricyanide to 900ml of water, will tend to give proportional cutting action; that is, the print highlights and shadow areas will be reduced at the same rate. Strong dilutions of potassium ferricyanide (30-to-60 grams to 900ml water) tend to reduce the print's highlight areas first. Stronger solutions, in conjunction with leaving the solution on the print for longer periods, will also tend to deepen the yellow stain. For normal usage, dilute the potassium ferricyanide to a pale yellow color. This is a reasonably weak solution. If one desires a more specific formulation, mix up Kodak Farmer's Reducer R-4A (one bath) for a strong cutting reducer: use Kodak Farmer's Reducer Formula R-4B (two bath) for a weak, proportional-working reducer. If one uses Farmer's Reducer in a one-bath system (the diluted potassium ferricyanide and hypo are mixed together) the solution will only keep for about 15 minutes. If one uses the two-bath system, which keeps longer, it is best to proceed slowly because image change is scarcely noticeable. Bleach in potassium ferricyanide for 15 seconds, fix, wash in water, bleach again for 15 seconds, fix, wash again, etc. Before beginning to bleach a print, soak it for a minute or so in water. After completion of either the one-bath or the two-bath system, follow up with normal print procedure: fix, wash, eliminate hypo, and re-wash. It is advisable to wear rubber gloves when working with bleaching compounds.

Remember, the reduction process continues for a few minutes after the print is removed from the bleach solution; so proceed cautiously. It is better to underbleach than to overreduce an image. Both solutions should be used at room temperature, although warming up of the potassium ferricyanide solution does tend to increase the rate of reduction.

2. Potassium Permanganate Reducer (Kodak's R-2 reducer formulation). This bleach tends to reduce the highlight areas of the print first, before affecting the darker print areas. It is a two-part stock solution: A. 14ml of sulfuric acid to 450ml of water (always pour acid to water); B. 24 grams of potassium permanganate to 450ml of water. To make the bleaching bath, mix 20ml of part A to 10ml of part B, plus 620ml of water. Using this formula, one can observe the bleaching action. But what is this? An obtrusive orange stain appears! Do not despair! A quick dip in fixer will contain the scoundrel. After bleaching, fix as in normal printmaking procedure, rinse, eliminate hypo, wash, etc.

3. Household Bleach Lastly, household bleach may also be used. It has a very sharp cutting action and should be diluted according to one's requirements. Use this bleach full strength as a time-saver for selective reduction of an image where complete silver removal is desired.

For additional information on bleaches, consult the Kodak Reference Handbook, section on "Chemicals and Formulas," or refer to the Photo-Lab-Index (Morgan & Morgan).

COLORING BLACK-AND-WHITE SILVER PRINTS

There are several methods for adding color to an image without employing full-color processing or color separation technique. Several alternatives are described below.

Hand Coloring Materials

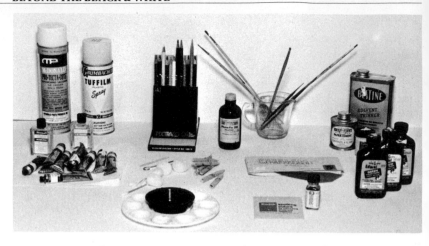

1. Commercial Toners. Certain manufacturers market toners ready to use, whereby one does not need to become immersed in any magical color chemical formulations. Toning can be carried out in room light. One starts with a fully developed, fixed, and washed print. Avoid fixers that contain hardeners. If the print has been dried, presoak in water for a few minutes. These toners share common characteristics and methods of application: **a.** Toners tend to soften the emulsion on a print. Therefore it is advisable to follow up most toning operations with a hardening bath. Kodak Rapid Fix has a separate hardener packaged with it. **b.** One may tone different portions of a print selectively by blocking out segments of the image with rubber cement (applied only to a dry print). To remove rubber cement, merely rub a finger across it and it will ball up. If problems are experienced, rubber cement thinner can help. Prethinning of rubber cement to an easier working consistency is advisable (mix 1:1, rubber cement to thinner). **c.** One can create a third color by toning in one transparent color and then another. Combine this method with selective toning technique for additional colors. **d.** Heat will increase the speed of toning. **e.** Toners can be applied by brush when tinting minute image areas. Use a brush that does not have a metal ferrule, or whose ferrule and adjoining bristles have been sealed with varnish. **f.** Stop ahead of time: toning will continue for a few minutes in the wash cycle. **g.** Different paper emulsions will give different toning results. **h.** Toners work best on papers that are free of hypo. Use a hypo eliminator after fixing prints that are to

Rubber cement resist method: (below left to right).
1. Apply a diluted rubber cement resist, with a brush, to those image areas not to be toned. 2. Tone the print. 3. Water rinse and remove the resist.

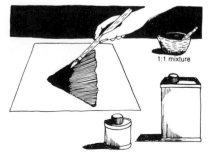

1:1 mixture

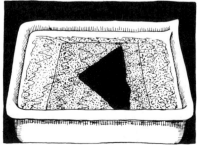

be toned. **i.** Soaking prints in water for five minutes before toning tends to give more even results.

Some of the toners one may wish to explore are: **a. Kodak Rapid Selenium Toner.** This toner does wonders to most straight black-and-white work. It adds a new dimension to the image, more depth and richer blacks. **b. Kodak Poly Toner.** This provides different tones from different mixtures. A 1:4 toner-to-water mixture gives a selenium tone and a 1:50 dilution gives a brown-sepia color. The sepia from Kodak Sepia Toner and the selenium effect from Kodak Rapid Selenium Toner are preferred by this writer; however, Kodak Poly Toner is easy to use, eliminating the need for stocking additional chemicals. **c. Kodak Sepia Toner.** With this product, the amount of tone attained is directly proportionate to the amount of bleaching of the print. Be CAREFUL of these chemicals: work in a well-ventilated room and wear gloves. **d. Edwal Color Toners.** Edwal offers five different colors: red, yellow, blue, brown, and green. Use them straight from the bottle with a ferrule-less brush or dilute them to make a toning bath. Again, prepare the brush for use by varnishing the ferrule and the bristles where they touch the metal. The intensity of these colors can be increased by adding 10ml of 28% acetic acid to every 900ml of working toner solution. Edwal-toned prints should not be refixed after toning. Avoid subjecting toned prints to direct sunlight.

Eastman Kodak and Edwal Scientific Products are only two of the companies manufacturing toners; other kinds are available. Many variables are involved in toning prints, such as brand and contrast grade of the papers used, degree of image development, and so on. Follow manufacturer's directions and few problems should arise. Beyond that, toning allows for experimentation, often with unpredictable results, some of which can be quite gratifying.

2. Coloring With Transparent Oil Colors. The Marshall Manufacturing Company is one of the oldest names in photo-oil colors. These transparent oil paints are very easy to use and come in a wide variety of colors. Be certain to order the extra-strong colors, which are more vivid, yet can be thinned to vary the shade and intensity of the color. As these paints are intended mainly for use on matte surfaces, glossy photos will first need to be given a "tooth" before the oils will hold properly. Try McDonald Pro-Tect-Cote Retouching Lacquer. Marshall also markets the necessary toothing lacquers and protective lacquer sprays.

Fine work can be done with transparent photo-oils that would be difficult to replicate with other coloring agents. When ordering, it is advisable to purchase the larger tubes of paint because of the substantial cost savings. There is, however, a starter set available. It includes 16 extra-strong colors in small tubes. Begin by ordering a set, or some basic red, blue, yellow, and black pigments from which other colors can be mixed as required. Many tints come premixed, however, and some individuals may prefer to start out initially with such stand-bys as Tree Green, Cheek, Lip, Flesh, etc. Include some Extender in your purchase, for use in increasing the transparency of colors (useful in differentiating highlights from shadow areas) and for rectifying errors. Include some Titanium White (artist's Flake

White), which can strengthen pastels as well as give colors an opaque quality. Also, order some of Marshall's Prepared Medium Solution for preapplication under paints and color pencils, as the colors may grip too strongly on some matte surfaces. Lastly, purchase a small bottle of Marlene, an oil-paint remover.

Photo-oil colors are best mixed on a palette or a plate and applied fairly heavily with little wads of cotton, using a circular motion, then rubbed down and blended with larger wads. Different-sized cotton tufts on the ends of toothpicks and skewers are very handy. A fine-pointed paintbrush, #000, will prove helpful for minute areas. The best hand-colored prints, no matter how varied the content, all share common elements of success. For example, the smaller details in the photo—such as eyes (including the whites), jewelry, quilts, etc.—are colored, adding a real vibrancy to the image. Also, be aware that colors selected do not have to correspond with reality in order to be successful.

Prints treated with sepia or selenium toner not only tend to be more permanent, but also take to coloring in a more pleasing way than untoned black-and-white prints do. An alternative to a separate toning step is to use a warm-tone paper, such as Agfa Portriga Rapid, which evolves brownish-black tones during normal development. Once the print has been oil colored and is thoroughly dry, it may be sprayed with a varnish or lacquer for added protection.

3. Watercolors, Food-Coloring Agents, and Inks. A black-and-white print may be toned selectively with a good-quality watercolor brush or tinted via a bath of photographic watercolor. Peerless Color Laboratories manufacture transparent watercolors designed for this purpose. These colors are packaged as a booklet of color-saturated sheets. Clip off a small piece of color and place it in a glass dish, adding enough water to obtain the desired working strength. Add a few drops of a wetting agent (e.g., Kodak Photo-Flo-200) or any good-quality dishwashing detergent (e.g., Ivory Liquid) to the working solution. This will assist color application, particularly on glossy prints.

There is a variety of other transparent watercolors, food-coloring agents, and inks (such as India inks) that can be employed successfully to tint photographs. Experimentation is encouraged. In addition, there are special pencils that release their color on contact with water.

4. Colored Pencils. Most colored pencils will color a photographic print. Experiment with different types. Mongol Water Color pencils can be either applied wet or dampened after application. Each method affords a different effect. The Marshall Manufacturing Company offers pencils specifically designed for photo application. As with the photo-oils, a "tooth" must be created on the surface of glossy prints in order to hold the color.

To spread pencil colors evenly once they are applied, use tightly rolled paper burnishers obtainable at art stores. A fine grade of sandpaper will help to keep pencil tips sharp. Pencils can be used in conjunction with oils when undertaking fine detail work.

Upon completion, spray with a protective lacquer or varnish.

5. Monochromatic Color Photographic Papers. Monochromatic color photographic papers produce black images on a colored ground. The paper is exposed to a black-and-white negative via enlargement and developed like any normal black-and-white print. High-contrast graphic images read well visually on these papers. There are several different brands available, such as Argenta and Luminos. Luminos colored paper is sold in standard photographic sheet sizes, as well as in rolls about 105cm (41 in) wide by about 31m (100 ft) long. It is packaged in a variety of colors: red, yellow, blue, green, gold, and silver.

There are also other types of colored photographic paper available for use with black-and-white negatives. For example, there are warm-tone papers, such as Agfa Portriga Rapid, that produce a brownish-and-white continuous-tone image instead of a black-and-white one. Another interesting type is Kodalith Ortho Paper. This is a high-contrast paper that is processed like Kodalith film. One can obtain light brown to dark brown and black tones. Other paper manufacturers market similar kinds of papers for the graphic arts.

6. Mirror-Finished and Metallic Black-and-White Prints. Rockland Halo-Chrome Mirror Developer will convert a silver bromide print to metallic silver, producing a black-on-silver image. Ferrotyping the print elicits a mirror finish. By using Rockland HC-2 Reducer in conjunction with Rockland HC Developer, one can obtain a silver-on-white print. The unexposed silver/gelatin areas can then be toned with Rockland colorants: red, green, blue, yellow. These toners create a similar effect to that of commercially made monochromatic colored papers.

Rockland also markets a photosensitive, enlargement-speed metal called Photo Aluminum. It renders a black-on-silver image and is available in various sheet sizes. It also can be toned with Rockland colorants: yellow produces a gold finish. All these materials can be handled under a red safelight and processed like a silver print.

7. Variations. There are many more possibilities for adding color to a black-and-white silver print: experiment with other coloring agents and methods. Try sewing colored outlines around photo images or airbrushing various watercolor pigments, drawing inks, and even very dilute solutions of acrylic paints onto matte photographic papers. Consider also the use of enamel paints, retouching dyes, crayons, glitter, vegetable juices, etc. As one can appreciate, the methods detailed here are only some of the easier and more controllable means of accomplishing that end.

THE DYE IMBIBITION PROCESS

The dye imbibition, or dye transfer, process need not be employed merely for the three-color reconstruction of a slide or an original scene. It may also be utilized to create posterizations from black-and-white tonal separations or employed for the selective addition of color and/or other colored images to a silver print. Once the silver image has been printed and processed (using a nonhardening fixer), the print can be mordanted to receive the color-utilizing dye transfer technique. Basically, this involves making a photographic matrix prepared via a wash-off relief process that enables dye to be absorbed in proportion to the thickness of the remaining gelatin layer representing the image, and to transfer this dye to the silver print.

A matrix is a special film coated with an unhardened gelatin emulsion. It has no antihalation backing; therefore, exposures can be made through the film base. Kodak markets both a pan matrix film (Kodak number 4149) and an ortho matrix film (Kodak number 4150). These films may be exposed by enlargement or by contact. Areas on the matrix may be dodged or burned, as with any conventional photographic paper. To make posterized matrices, run a test strip to establish normal exposure and then vary the matrix exposures accordingly.

Accurate registration of tonally separated matrices during printing requires a systematic procedure of exposure. See Chapter 8, section on color separations. Prior to exposing the matrices, place the processed silver print over two registration pins centered on the sides of the enlarger's baseboard and tape the sides of the print down flat. The print should have been made with a border so than an unexposed border will surround the matrix to facilitate handling. The negative used to make the print is inserted into the enlarger and projected onto the print in such a way that the printed and projected images are aligned. Place a sheet of matrix film face down over the registration pins: it is face down when the Kodak notching system is at the upper right-hand side. Code each matrix and record the exposure.

The matrix is processed emulsion-side up, using either Kodak Tanning Developer A and B or a hardening bleach method. The bleach method, or wash-off relief technique, is outlined below:

1. After the matrix film has been exposed, develop it in Kodak DK-50 undiluted, or in HC-110 diluted 1:4, for three to five minutes at room temperature with agitation every 30 seconds.

2. Wash the matrix for approximately four minutes in several changes of water at 20° C (68° F).

3. Bleach the matrix in Kodak R1OA Hardening Bleach for three to five minutes. This bleach consists of two parts:
   A: 10 grams ammonium dichromate
      2ml sulfuric acid
      500ml water (pour acid slowly into water, not water into acid).
   B: 23 grams sodium chloride (kitchen salt)
      500ml water
Make a working solution by combining one part each of A and B to six parts water at 21° C (70° F).

4. The unhardened, unexposed areas of gelatin corresponding to the black areas on the negative are now washed off in hot water (38° to 45° C; 100° to 120° F). Do not subject the matrix to direct water pressure, but handle it with extra care because it is particularly susceptible to damage under hot water. Room lights may be turned on for this and the remaining steps.

5. Once the white milky matter has been completely removed, the remaining gelatinized image is set by chilling it in water (about 19° C; 65° F) for three minutes.

6. The matrix is now fixed in a nonhardening hypo solution for

five minutes. Use Kodak Rapid Fix (diluted as for film) minus the hardener.

7. Wash the matrix again for five minutes at 21° C (70° F) and hang it to dry, suspending all matrices from the *same* corner in relation to the image. If one desires to remove portions of the matrix selectively, rewet the surface with water and use an X-Acto knife to scrape off unwanted areas.

Many black-and-white photographic papers can be mordanted successfully for dye transfer. Avoid those that have rough or matte surfaces. The function of the mordant is to make the gelatin receptive to the dyes, which have an affinity for the aluminum salts in the mordant. One useful mordant is Kodak Formula M-1. It consists of two solutions:

A: 20 grams sodium carbonate, anhydrous
     250ml water.
B: 100 grams aluminum sulphate
     500ml water.

Slowly add Solution A to Solution B, stirring with a glass rod until the white effervesence subsides. The processed print is mordanted in the above solution at 20° C (68° F) for approximately five minutes and then washed for four to six minutes. Next, treat the paper in a 5% sodium acetate solution (29 grams per 600ml water) for five minutes and rewash for approximately ten minutes. The print may be dried and stored indefinitely, or used immediately. Prior to dye transfer, place the print in a solution of Kodak Dye Transfer Paper Conditioner for at least 15 minutes. It can be left in this solution for several hours.

The matrix is prepared for transfer by staining it with dye from a Kodak Matrix Dye Set. Use only distilled water when mixing dye. Colors can be premixed and stored in glass containers. Color contrast is increased or decreased by altering the amount of buffer added. Presoak the matrix in hot water (43° C; 110° F) for two minutes, then either build up color selectively with a soft brush or submerge the matrix, face up, in a dye bath for five minutes. Agitate the matrix periodically. Once the matrix has been sufficiently stained, place it in a 1% acetic acid bath for about one minute (10ml glacial acetic acid to one liter of water). Transfer the matrix again to a holding bath (5ml glacial acetic acid to a liter of water) until it is ready to be used. These baths eliminate unabsorbed dye prior to transfer.

The mordanted silver print floating in the conditioner is now placed over a pair of registration pins that are secured to the work table. Use a wide squeegee to remove excess conditioner from the surface of the paper. Take the dye transfer matrix out of the holding bath, place it over the registration pins, and lay it down on the print so that no air will be entrapped. It is extremely important that the matrix still have some moisture from the holding bath on it to promote even contact. Next, roll a soft rubber roller over the back of the matrix. Transfer time will vary with the temperature and the dye: six or seven minutes usually is enough. Once the transfer operation is complete, rinse the matrix in Kodak Matrix Clearing Bath CB-5 to remove residual dye stain. The matrix may now be used over again with another color. As dye colors have a tendency to bleed, it is best to dry the print as soon as possible after each transfer, in order to minimize this problem.

LITHO-ETCH/DYE-STAINING OF HIGH CONTRAST FILM

The litho-etch/dye-staining process is a controllable means for selective addition of brilliant color to a high contrast lithographic transparency. The technnique involves the removal of processed silver/gelatin from the film base and the staining of the remaining gelatin with dye, thus creating a colored transparency. If a negative transparency is etched in this way, a colored positive image is obtained after dye application. The manipulated transparency can be used as a color slide, combined with other colored transparencies for poster effects, employed as a color overlay, or used as a dye matrix to transfer the image to photographic paper. The process can even be employed as an image-reversal technique. The procedure is as follows:

1. Expose and Process the Tranparency. Fully expose and develop a sheet of high-contrast orthochromatic film, such as Kodak's Kodalith Ortho Film Type 3 or 3M's Line Ortho PL-5, so that the image is processed to film base. This insures that all the gelatin in close association with the exposed silver will be removed and therefore incapable of becoming unintentionally stained with the dye. Process the transparency in Kodalith A and B developer, following standard procedure. After development, the film is rinsed in water, fixed, washed, and dried. If it is desired that the image be reversed, wash the film briefly in warm water immediately after the development bath, then place it in the etching solution (see next step). Once all the exposed image has disappeared, heavily re-expose the transparency to strong white light to fog the remaining silver/gelatin. Redevelop the film, fix normally, and wash in running water. Whenever a continuous tone-like image is desired, use halfton on texture screen (e.g., mezzotint) when making the transparency (see Chapter 8).

2. Etch the Transparency. Once the film has been processed, it may be stored indefinitely before being etched. In this respect, it is handled like any high contrast film. To etch the transparency, place it in the etching solution. One can actually see the exposed image or blackened emulsion dissolve away. Keep the film in the etch, emulsion side up, until only a milky white gelatin relief remains. Agitate the film periodically. Gently apply a cotton swab to assist the removal of any stubborn emulsion. Etching time will vary with the strength and temperature of the solution, and may take from 5 to 20 minutes. After the transparency has been etched, rinse it in running water for 2½ minutes. The etched side of the transparency is quickly ascertained by running a finger lightly across the film surface. A slight relief will be evident.

The etching bath consists of two parts:

Stock Solution A: 25 grams citric acid
20 grams copper sulfate
8ml nitric acid (optional: tends to speed up etching process)

240ml hot water (about 59° C; 135° F)
Stock Solution B: 3% hydrogen peroxide (normal medicinal
strength). About 300ml will be needed.

Mix equal parts of stock solutions A and B just prior to use. Set the
tray of etching bath in a larger tray of hot water during processing, to
maintain the higher temperature of the solution. In order to curtail
early contamination of the etching bath, periodically remove the
transparency from the etch and place it in a hot-water bath for 15
seconds. This seems to promote removal of the blackened silver
residue. As the etchant loses its potency, replenish the bath with
freshly mixed solution and recheck the temperature.

3. Prepare the Etched Transparency to Receive Dye. Before staining
an etched transparency, place it, relief side up, in an acetic acid
holding bath made by mixing 5ml Kodak Indicator Stop Bath into
460ml water. An alternative to the holding bath is to wipe the
transparency with 28% acetic acid prior to dye application.

4. Apply Color to the Transparency. Color for staining the transparency may take the form of an extremely fine powder or else of a
suitable liquid substance. A variety of pigments can be used, including enameling paints, fabric dyes (e.g., Procion), watercolors, photographic dyes, etc.

Liquid dye is applied to the etched transparency by loading a
brush and layering several coats in order to intensify the color of the
image. Powders can be applied by rubbing the dry pigment gently
about the surface. Avoid getting any color on the slightly gelatinized
back side of the transparency, as it will stain. This potential problem
can be minimized by taping the transparency down on a sheet of
glass, so that all the edges are sealed during color application. A
(¼-inch) .5cm clear edge around the transparency will also facilitate
handling.

As the staining progresses, the transparency can be returned
periodically to the holding bath to remove dye from the nongelatin-
ized relief areas. This makes it easier to observe color intensifica-
tions. Change the bath as the dye discharge accumulates. In order to
remove unwanted dye from gelatin-stained portions, place the film
in a hot-water bath. If this doesn't clear the film successfully, return
the transparency to the etching bath for a few minutes.

The dyes in the Kodak Matrix Dye Set work well for transferring
the image to processed photographic paper. (Try Agfa's Brovira
black-and-white paper.) Before staining, soak the paper receptor in
water and acidify slightly with acetic acid in order to promote a
better transfer. Remember that the thickness of the gelatin coating on
a lithographic film will prohibit dye transfers as color-intense as
might be obtained using Kodak Matrix Films. The image will reverse
during transfer; so plan accordingly. A review of the dye imbibition
process will prove useful in perfecting technique.

TRANSFER TECHNIQUES

There are several ways to present a silver image other than by
conventional dry-mounting. Without having to use special emulsions, one can transfer a processed photographic image to a variety
of surfaces. There are several products available: Lift-A-Print, by the

Artistic Manufacturing Company; or Decalon, The Instant Decal Medium, by Sangray Corporation. These are decalomania materials and can be found at variety stores, sewing centers, hobby shops, etc. The material consists of an adhesive-backed sheet of special film, a decal medium. When the image is placed in contact with the film, the image adheres and becomes transferred to it. The image can then be transferred again, but this time permanently, to just about any surface imaginable.

Another interesting method entails laminating technique (Ademco Heat Seal Films, Photo Technical Products). With this method, one can make oversized laminated photo transparencies, laminated photographic table mats, interior photodecorated and laminated wall panels, transfers of photographs to canvas, etc. For canvas surfaces, a resin-coated photographic paper must be used for the transfer. One can also give the surface a glossy or a matte look. Unfortunately, this technique requires pressure and heat from a special dry-mount press and heat-seal films. Both equipment and materials are expensive, but results look very professional.

A less expensive method for transferring the continuous-tone image from a resin-coated photo paper involves first soaking the processed print in very warm water. Next, rub and peel off the back layer of polyethylene film. The image remains on a flimsy piece of resin. In this state the film tends to curl severely, but it can still be used in this flexible state by bonding it to some surface. The chosen support can be transparent or opaque, depending upon the effect desired. A variation of the above technique is offered by McDonald Photo Products, Inc. It involves a laminating-and-gluing procedure recommended for use with their products when mounting a resin-coated image.

Silver prints may also be mounted by more common découpage techniques. Thin, single-weight paper (not resin surfaced) should be used, as it gives the best appearance. First, the image is glued (with Elmer's Glue All or framer's glue) to the desired support—wood, metal, rock, etc. Then it is given five or six coatings of clear plastic resin, being allowed to dry between coats. The resin will protect the image from damage. Although this method is very simple, its major drawback is that the image looks like a photograph under plastic rather than like an integral element of the article it is attached to.

Kent W. Wade, "Anna at 87"... Resin-coated photograph, Heat/Pressure Laminated onto canvas.

Chris Nelson. "The Yamaha F-160"... B & W images of a guitar were glued onto a hand-carved piece of cedar creating a 3-dimensional sculpture. A fine example of decoupage mounting, where the photographs look like photographs mounted on wood, yet enhancing image presentation. 5"L x 2½"W x ¾"H

Non-Silver Printing Processes Re-examined

Examination of the historical development of photography reveals a multitude of abandoned, yet interesting processes and techniques that are now being re-examined. Many of the old formulas utilized light-sensitive iron salts or chemically sensitized, pigmented colloids to produce a photographic image. Unlike the mass-produced silver papers of today, a wide variety of surface options and sensitized coatings was available. These nonsilver emulsions were slower, of course, and had to be prepared from the raw materials, requiring the individual to be both photographer and chemist. However, from these early darkrooms came some incredible results unmatched even by contemporary wizardry. This chapter deals with some of the easier, less expensive, and more visually stimulating nonsilver printing processes.

THE CYANOTYPE, A Printing Process Utilizing Ferric Salts

Different salts have been experimented with because of their light-sensitive properties, such as cobalt and cerium salts. None has been as fully explored nor appears to have had as great an impact on photography, however, as the iron or ferric salts. The various reactions of ferric salts, when combined with other chemicals and exposed to ultraviolet light, have led to several different approaches to image-making. Some processes rely on the hardening effect of ferric salts on a colloidal compound, such as ferric chloride on a gelatin colloid. This might be compared to the gum bichromate process (to be discussed). Other processes rely on the ability of ferric crystals to react chemically with certain metallic compounds to form images of metal precipitate; for example, the platinum, palladium, or kallitype processes. There are also, however, techniques that rely on the reaction of ferric salts with certain chemical compounds to form a colored by-product that creates the image. This knowledge has been applied in a number of processes, of which the cyanotype is the best known.

The cyanotype, often referred to as the "blueprint" or "ferro-prussiate" process, is a negative/positive process that yields a blue image on a white ground from a negative transparency. Normally the paper should be sized before being sensitized. This is not always necessary and will depend upon the surface of the material employed. Most of the heavier-weight watercolor papers can be used without treatment. Sizing will help to prevent the light-sensitive salts from soaking too far into very absorbent papers, which would result in an excessive development time. It also helps to minimize yellow staining of the highlight areas due to the potassium ferricyanide.

There are many different substances and methods that can be used to size paper (gelatin, albumin, etc.). The simplest to use is a household spray starch. Apply a light coat in a crisscross fashion, then gently rub the starch over and into the surface of the paper with a damp sponge. Starch may also be purchased in powdered form: Boil a small quantity in water, to obtain a thin consistency, then apply it to the paper while the solution is still warm.

It is very easy to make a cyanotype sensitizer. Two stock solutions are necessary:

A: 25 grams ferric ammonium citrate (brown crystals)
    125 ml distilled water
B: 17 grams potassium ferricyanide (not ferro)
    125 ml distilled water.

Once these solutions are prepared, they may be stored in separate opaque containers indefinitely. It is preferable that solution B be prepared at least one day prior to use, to be sure that it is completely dissolved.

When ready to coat the paper, mix equal parts of A and B. The mixture will remain useful for the length of a normal working session. If one desires a lighter, more delicate shade of blue, dilute the working solution by adding more water. To deepen it, see note on washing, below. To sensitize the paper either soak it for two minutes or apply the solution with a soft brush. Brush strokes should be from

Materials for the Cyanotype Process.

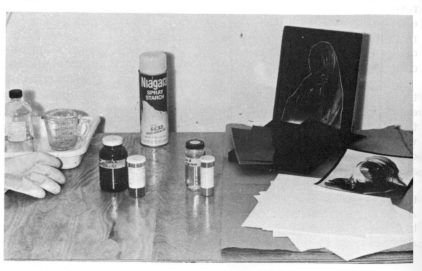

top to bottom of the paper and from side to side, to insure an even coating. Dry the paper on a flat surface or hang it up, rotating the top to the bottom periodically to prevent streaks. Sensitizing through development should be done under subdued light. After the paper is dry, it may be used immediately or stored in a dark place temporarily. Sensitized paper has been successfully exposed 96 hours later.

To expose the paper, place a negative transparency (continuous tone images made with Kodak Super XX Pan sheet film work well) in contact with the sensitized paper and sandwich it between glass or use a contact-printing frame (see below). Any ultraviolet light source will work—sun, sun lamp, what have you. A test strip will help determine the best length of time. The emulsion is similar in response to that of printing-out paper: you will be able to see changes in the paper as the iron salts react with the light. Rather than make a separate test strip, place a negative, pennies, etc., on a corner of the glass/paper sandwich and monitor changes in the paper's discoloration by periodically lifting the object off.

The best type of contact-printing frame to use is one that has a tensioned split back, permitting periodic inspection of the transparency/paper sandwich without misaligning them. A major benefit of this device is that over-all contact between paper and transparency will be better distributed than when using just a sheet of glass. Of course, a vacuum frame is the ultimate system. Contact frames are available through photographic dealers.

Once the paper has been exposed, develop the image by rinsing the paper in several changes of fresh water for about 20 minutes. Adding a few drops of hydrochloric acid to the initial rinse (5 drops to 1 liter of water) will deepen the blue considerably. **CAUTION: Any acid that comes in contact with any cyanide will form deadly,**

The Cyanotype Process: (left to right from top) **1.** *Soak paper in equal parts of sensitizing solution for 2 to 3 minutes and hang to dry.* **2.** *Expose paper to a negative transparency with an ultraviolet light.* **3.** *Contact printing frames are handy for periodically checking exposure progress without misaligning the transparency.* **4.** *Develop the image in a water bath to which a few drops of hydrochloric acid have been added. This will deepen the blue color significantly.* **5.** *Continue to develop/wash the paper in several changes of tap water until all traces of unexposed chemical is removed.* **6.** *The completed cyanotype. The transparency used was made with Kodak Super XX pan film to maintain a longer tonal gradation. The enlarged negative was made from a 35mm slide.*

poisonous cyanide gas. Be certain, therefore, that the cyanide solution is stored away before you open the acid. A weak ammonium dichromate or a 3% hydrogen peroxide solution will also deepen the blue color. Run a test to determine which intensifier is preferred. As water also serves as the "fixer," the print may now be hung to dry.

Cyanotypes may be toned to achieve some color variation. To make a brownish-black color, immerse the processed cyanotype in a solution containing 10ml of 18% ammonium hydroxide plus 100ml of distilled water; once the blue color has been completely removed, wash the print and place it in a solution of 10ml tannic acid to 500ml distilled water. A greenish tone is also possible: After washing, place the print in a weak 1% solution of sulfuric acid. To obtain a violet tinge, try immersing the cyanotype in a cold borax solution.

A multitude of surfaces will accept the cyanotype sensitizer: fabric, leather, bisque ware, etc. In addition, color can easily be integrated into a cyanotype image using contemporary hand-coloring techniques, toning methods, or colored paper stock.

One very effective way to present a cyanotype is with a thin, white strip around the entire image to separate it from the larger, blue border area. Make a mask of opaque tape (litho tape or 3M silver Mylar tape #850) cut to fit around the edges of the negative during exposure. Make the mask at least 6mm (¼in) in width. The area under the mask will not be exposed and will print white.

Another visually interesting presentation of a cyanotype is to give it the appearance of an etching. Etched prints or engravings have a sunken, or impressed, image the size and thickness of the metal plate that was pressed into the paper during printing. If one is using a thick watercolor or printmaker's paper, this effect can be simulated by making a cardboard template the exact size of the image area. For example a 20x25cm (8x10in) image, although printed on larger paper, would require a 20x25cm (8x10in) cardboard template. Soak the paper until it is moist and pliable. Place the template over the image, aligning it perfectly so that it just covers the image. Tape the template lightly in place and carefully invert the paper/template combination. Now, rub a burnisher around the edges of the cardboard, working the paper down over them. (Any blunt object will serve—toothbrush handle, spoon, etc.). When the paper is turned face up and the template removed, the image will appear to be recessed into the paper as if by the action of a printing press. A multiplicity of different template shapes and sizes can be made to match the design of the transparency one intends to use. Several transparencies and templates may also be combined on a single sheet of paper, creating a multiengraved-print effect.

THE DIAZO PROCESS, An Alternative to the Cyanotype

The cyanotype is only one of several more complicated iron salt processes. There is the pellet process, which produces a blue image using a positive transparency; the ferric-gallic process, which produces a black image on a white ground, and the ferric-cupric process, which yields red and purple tones. Another, simpler, alternative also exists for creating ferric-salt-like images: the diazo process. Paper treated with a light-sensitive diazonium salt and a coupler

Making a Cardboard Template Etched-Like Print:
Above: Presoak the print for 15 minutes.
Below: Tape the cardboard template over the image area and burnish the backside to recess the image.

compound is exposed to ultraviolet light, which decomposes the salts in areas corresponding to the clear areas of the transparency, making a positive image from a positive transparency. The paper is processed in ammonia fumes. This development makes the sensitive coating alkaline, which causes the coupler and the unexposed diazo crystals to link up, forming the dyeline image.

Diazo-treated paper may be purchased very inexpensively from architectural and engineering supply firms. This paper has virtually replaced blueprint paper because it is less expensive and requires minimal processing time. Diazo paper is available in a variety of sheet and roll sizes and in three weights: standard (20#), heavy (24#), and cardstock (32#). It comes in blueline, redline, brownline, or blackline; that is, the line (the image) is in one color on a white ground. There are several speeds of diazo paper: slower speeds give more control, contrast, and deeper shadow areas. Experiment with No. 5 speed brown or black line paper for optimum quality rendition under studio conditions. If one wants to use more exotic diazo materials, there are also both transparent and translucent sensitized Mylar surfaces available. These surfaces require heat for processing, but this can be applied by blowing a hair dryer over the outside of the development tube (described below). Lastly, Scott Graphic markets monochromomatic/transparent Diazochrome film on a plastic base. These materials are normally employed in color proofing, although they do lend themselves to creative manipulation.

Diazo-treated paper is an excellent surface on which to draw, paint, or hand-color with transparent ink markers. It can also be cut easily if one wants to incorporate processed segments into multiple imagery. The major drawback is that diazo paper will fade eventually if exposed for prolonged periods to UV light.

Diazo paper requires a positive transparency in order to obtain a positive image. It is an ideal material for making extremely large, yet very inexpensive, photograms (shadowgrams).

When making photograms, try using a portable UV light source, moving it around the object, creating high and low "relief" by making shadows of different densities. The paper has a sensitive yellow coating on it that "burns away" when subjected to UV light [for instance: one to five minutes with a sun lamp at 63.5cm (25in)]. The remaining yellow is processed in ammonia fumes to give it color. Use concentrated ammonium hydroxide (26° Baumé). **BE CAREFUL: This chemical is hazardous!** Ordinary household ammonia may also be used, although longer development times will be necessary. Once the image has been exposed, roll up the paper and place it in a developing tube (see drawing). Pour a small amount of undiluted ammonia into a glass dish and lay a piece of pegboard over it. Next, place one end of the tube over the pegboard and seal off all holes not under the tube. Seal the opposite end of the tube with a piece of cardboard. The ammonia vapors will rise and be entrapped in the tube, enveloping the exposed paper. In approximately three minutes the paper can be removed, completely developed. No additional processing is required.

Diazo prints are simple to make and require little equipment. If the need arises, print shops have diazo machines that will process

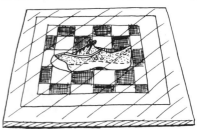

The Diazo Process:
Positive/Positive Process
Above: Expose the Diazo paper to a positive transparency or object. Below: Develop the image with ammonia fumes. Heat is sometimes applied to the outside of the tube with a hair dryer in order to assist development.

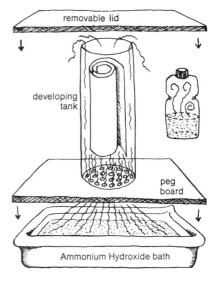

Colloid Rendered Insoluble: Negative/Positive Process: (left to right) **1.** *Pigment premixed into colloid and coated on a piece of paper. Paper exposed to a negative transparency, both sandwiched together under a sheet of glass.* **2.** *Water development: unexposed image areas dissolve in the water—exposed areas are hardened by the light.* **3.** *Dry the print. (Optional: clear dichromate stain and harden colloid in alum bath.)*

images up to 107cm (42in) wide by any length. This could be useful for making wallpaper murals, which would be limited only by one's imagination and the size of one's dwelling. It would be necessary first to have a printer enlarge the image onto super chunks of litho film (91x107cm; 36x42in). These would then be spliced together with tape for the diazo exposure.

PRINTING PROCESSES UTILIZING DICHROMATE-SENSITIZED COLLOIDS: Theory and Alternatives

There have been many and varied processes over the years that utilized the basic principle that light will act on certain organic substances, such as colloids, when they are mixed with a dichromate solution. In fact, many of these processes are used today in photolithography, photosilkscreen, photoetching, and other techniques.

 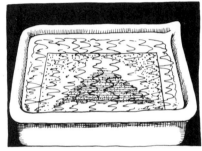 

What is a colloid? It is an organic substance soluble in water, such as dried albumin (egg white), gelatin, fish glue, caramel (sugar), and gum arabic (from trees); other substances such as Elmer's School Glue or polyvinyl-alcohol/polyvinyl-acetate solutions may also be employed. When the colloid is mixed with a dichromate it becomes light-sensitive and takes on some very important characteristics: certain soluble colloid mixtures can be rendered insoluble when exposed to UV light; when exposed to UV light, certain other colloids will lose their "sticky" quality; lastly, some colloid mixtures when exposed to UV light will become hard, while the unexposed areas will swell up and absorb water.

The dichromate solution used to sensitize the colloid can be made of either potassium dichromate or ammonium dichromate crystals. The ammonium salts are more sensitive to light, and have printing times twice as fast as those obtained using potassium salts. The salts may be purchased as ammonium bichromate or dichromate (the terms are now interchangeable). A saturated solution (one part ammonium dichromate to five parts water) is usually desired.

Gum bichromate printing is one of the photographic processes based on the phenomena that a sensitized colloid sandwiched with a negative transparency and exposed to UV light will yield a positive image upon water "development." The light hardens the emulsion where it strikes, while the areas covered by the negative, and therefore less or unexposed, will, depending on their degree of exposure, be dissolved out in the water. If the colloid has been pigmented with

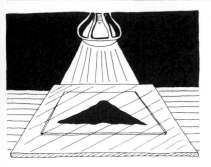  

a water-based ink, the image will appear in color and will give an almost three-dimensional relief effect.

A process (and derivation from it) that utilizes the "sticky" quality of the sensitized colloid is referred to as the "dusting-on," or "powder" process. With this process, a positive transparency is used to obtain a positive image. When a sugary gum dichromate solution is coated onto a substratum, dried, and then sandwiched with a transparency under glass, the light will harden the surface and cause it to lose its tackiness. The remaining, unexposed, surface area will possess a tacky quality varying in stickiness according to the amount of UV light that struck it. By dusting on a finely ground powder (such as carbon black), one can make the image appear. This technique can also be employed in the fabrication of ceramic and enamel decals as well as in the etching of glass.

Lastly, iron salts, such as ferric citrate, also were utilized as sensitizers prior to the discovery of dichromate. Paper thus treated would be dusted over, creating a powdered picture.

Rather than mix the pigment directly into the dichromated colloid before exposure, some processess (e.g., collotype) rely on the swelling action of the emulsion as the means to produce the image. In these instances, an unpigmented dichromated gelatin is used to coat a piece of paper. A negative transparency is required to produce a positive image. The transparency sandwiched between glass and the sensitized paper is then exposed to ultraviolet light (sun, photoflood, sun lamp, etc.). Where the light is able to strike the surface of the sensitized paper, the emulsion will lose its ability to absorb moisture. After exposure, the paper is soaked in cold water at 15.6°C (60°F) until the nonimage areas swell up. Next, the paper is brushed or rolled lightly with a greasy lithographic ink in the color of one's choosing. Those unexposed areas that are water-saturated will repel the ink, while the light-hardened areas will accept the pigment. Similarities between this process and contemporary lithographic processes are apparent.

One adaptation of this technique utilizes the swollen gelatin in an interesting way: The inked image is covered with plaster of Paris, so that a relief image is cast in the plaster from the swollen paper. Another derivation uses a positive/positive system: certain inks and dyes are added to the water and absorbed into the receptive, unexposed areas; these dyes are "special" in that they leave the light-hardened areas unaffected. Many processes have relied upon the above principles, such as the Bromoil and the Oil processes.

Tacky Colloids: Positive/Positive Process: (left to right) **1.** *A non-pigmented colloid is coated onto paper and exposed to a positive transparency. Both are sandwiched together under a sheet of glass.* **2.** *The image is developed by sifting or dusting-on an ultra-fine powder which adheres to the tacky, unexposed areas.* **3.** *A fine brush is used to "push" the powder about the image area increasing the intensity of the image by adding more layers of powder. Intermittently, moist air is gently blown onto the surface of the image area to make the unexposed portions more receptive.*

 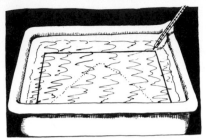 

Swollen Colloids:
Negative/Positive Process: (left to right) **1.** *A non-pigmented gelatin colloid is coated onto paper and exposed to a negative transparency. Both are sandwiched together under a piece of glass.* **2.** *The image is prepared for development by soaking it in cold water (60°F). The unhardened areas swell up with water.*
3. *Develop the image with a suitable ink. The light-hardened areas accept the ink, while the swollen areas repel it.*

Following are some very useful facts regarding the mixture, application technique, and exposure variables typical of most dichromated colloidal compounds:

a. Additional quantities of ammonium dichromate will increase the responsiveness of the colloid to ultraviolet light up to the point where dichromate crystals begin to form on the surface of the colloid. Mixtures range from a useful compromise of 1 part ammonium dichromate solution to 3 parts colloid up to 1:20 mixtures. Even these small quantities will harden to colloid; however, exposures will tend to be long.

b. The thickness of emulsion coatings should be minimal, because light must penetrate all the way through the coating to harden the colloid properly and too thick a coating will necessitate excessively long exposures. If the coating is overly thick, the light may not be able to penetrate at all, thus totally eliminating any possible image formation during development.

c. The ammonium dichromate (sensitizer) and the colloid compound will last indefinitely, if stored separately. Once mixed together, however, a hardening action takes place immediately, even in the dark. Thus, coated paper should be exposed soon after it is dry for optimum, controllable results.

d. Image contrast decreases with increases in the ratio of dichromate to colloid, although sensitivity of course increases. If the intention is to explore the potential of the gum bichromate process fully, it would be useful to mix several batches of varying contrasts for use with negatives of varied densities and exposures.

e. Finally, humidity and moisture do tend to increase the sensitivity of a dichromated mixture. Reason enough to move to the New Hebrides!!!

GUM BICHROMATE PRINTING: Contemporary Technique, Materials, and Application

Gum printing is a simple, controllable, and inexpensive process which can be done under subdued light. It yields beautiful, rich-colored images and has a wide variety of applications. The name of the process originates from the use of gum arabic as one of the major ingredients—the colloid base component. Gum arabic is an organic gummy substance that comes from the stems of the acacia tree found in different parts of Asia and Africa. It is a hygroscopic substance, one which tends to absorb water or to draw moisture to itself. It also has an adhesive quality, and it is affected by ultraviolet light when mixed with a dichromate solution.

Gum arabic is available in crystal or in powder form. A stock solution (part A) is made by mixing 28 grams of gum arabic with 90ml of hot water, but below the boiling point. It is not necessary to use a high grade of gum; one can obtain a satisfactory quality from printmaking and ceramic-ingredient suppliers. As a less refined grade will contain impurities, strain the mixture through several layers of cheesecloth. A useful gum colloid may also be obtained in a liquid state from a lithographic supplier as 14° Baumé gum. This solution is packaged ready for use in a consistent nonsouring form.

Gum arabic does not have very good keeping qualities; therefore, prepare only enough for a few days at a time, unless steps are taken to preserve it: refrigeration definitely will help, as will three to five drops of carbolic acid (phenol) to the batch. These measures will prolong the life of the solution up to several months: the acid acts as an antiseptic and eliminates unwanted bacteria that would cause the gum to go foul. Let the gum solution (part A) set for at least 24 hours before using: mix it the day before a work session. Gum viscosity can vary with age and by batch; so strive for an average working consistency based on the dilution given above.

In addition to the gum arabic solution (A), a sensitizing solution (part B) of ammonium dichromate, is required for printing. Mix 28 grams of ammonium dichromate into 150ml water at 37.8°C (100°F). Cool to room temperature before use. It may be prepared as required or kept in solution. It is advisable to wear rubber gloves when working with this chemical, as it is possible to get chromate poisoning. Breathing the chemical isn't recommended, either.

To make a sensitized gum solution for printing, the following procedure is offered: Mix one part gum solution (A) to one.part sensitizing solution (B); but before combining them, add the desired pigment or color to the gum solution (A). Recycled 2-ounce plastic pill containers with snap-on lids are ideal for mixing purposes. Mark off 15ml (½oz) and 30ml (1oz) on the outside of the vial. Pour in 15ml of gum solution (A) plus the pigment and stir. Next, shake the container 50 times. (It is important that the pigment be thoroughly mixed with the gum.) Just prior to coating the selected surface, add

Gum bichromate printing materials for 3-color separation from a color transparency.

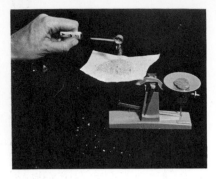

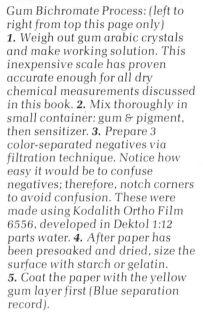

Gum Bichromate Process: (left to right from top this page only)
1. *Weigh out gum arabic crystals and make working solution. This inexpensive scale has proven accurate enough for all dry chemical measurements discussed in this book.* ***2.*** *Mix thoroughly in small container: gum & pigment, then sensitizer.* ***3.*** *Prepare 3 color-separated negatives via filtration technique. Notice how easy it would be to confuse negatives; therefore, notch corners to avoid confusion. These were made using Kodalith Ortho Film 6556, developed in Dektol 1:12 parts water.* ***4.*** *After paper has been presoaked and dried, size the surface with starch or gelatin.* ***5.*** *Coat the paper with the yellow gum layer first (Blue separation record).*

15ml of sensitizer (B) and shake again 50 times. Try to do all mixing of the emulsion at room temperature. Use the better-quality water-based colors with their richer, deeper pigment saturation. Shiva water-based woodblock inks are versatile and give very pleasing results, as do Windsor & Newton watercolors. Colors can overlap and mix to create a new hue. When doing multiple coatings, it is best to plan a color scheme in advance. In order to determine how much pigment is required for any given color, keep a chart of the quantity used. Keep track by the length of the "worm" emitted from the tube. For example, a half-inch (13mm) worm of Shiva to 30ml of sensitized gum solution (part A plus part B) will generally give a deep, rich color. Dry powder pigments may also be used: measure by weight the quantity employed.

As the gum-plus-pigment combination is blended with the sensitizer, one will notice a color change. Do not be alarmed. This is due to the inherent color of the ammonium dichromate crystals (orange) and can be cleared later with an alum hardening bath. With the emulsion now mixed, the paper is ready to be coated. As with the cyanotype process, it is advisable to use a good quality of paper that will withstand repeated soaking and drying, such as Rives BFK, Arches Silkscreen, Strathmore Bristol (two-ply), etc. Also try using Aquabee watercolor paper. Its textured surface tends to break up the image slightly, so that sharpness forfeited when enlarging small negatives to contact-printing size is not so evident. This texture also seems to reinforce the three-dimensional feeling of the print when multipigmented coatings are required.

If you plan to do multiple colors, or if one is using an unsized paper, then the paper should be sized prior to application of the

Gum Process continued: (this page, left to right from top) **6.** Dry the paper in subdued light. **7.** Take blue separation and outline the transparency edges with pen. **8.** Expose the blue separation record/coated paper sandwich to an ultraviolet light source. **9.** Develop the image in water at 70°F. Use a brush to gently remove unexposed areas. **10.** Realign the 2nd separation (green record) under the blue separation on a light box and tape together. Recoat the 2nd color (magenta) and align the blue separation between original pen lines on the paper. Tape down the green separation and remove the top (blue) separation for the exposure. Repeat procedure with 3rd color separation.

emulsion. Starch is useful for sizing paper. It helps to prevent bleeding of colors into each other as well as into the base stock. Before sizing, soak the paper in hot water for about 12 minutes and dry it. This preshrinks the paper, which is essential if one intends to do multiple layers of color and expect reasonable re-registration. Of the several ways to size the paper, the easiest is to use ordinary spray starch. Commence application by spraying at the bottom in horizontal sweeps across the paper, working toward the top. Be careful not to put on too much starch or else the emulsion might not adhere. Use a sponge to wipe off the bubbling excess and to insure even distribution. When using spray starch, a thin coat should also be applied between each layer of a multicolored print.

A more exotic sizing technique for one-time application consists of mixing 15 grams of supermarket-grade gelatin into 700ml cold water. Once the gelatin has absorbed most of the water, heat the solution until it dissolves thoroughly. While the solution is still hot (46°C;115°F), apply it to the paper with a brush or by soaking. After application of the sizing, gently pull the paper between two glass rods in order to eliminate excess solution or bubbles. Two coats may be necessary, drying between coats with a hair dryer or fan. Next, harden the surface by floating the paper in a solution of 10ml of 37% formaldehyde to 350ml of water for about two minutes and dry the paper. The formalin hardens the gelatin protein. Beware of the fumes during this hardening process.

To apply the emulsion to the sized paper, a 5cm to 8cm (2 to 3 in) soft brush is recommended. Tape the paper down to the backside of a tray larger than the paper size to facilitate coating and handling. Work the brush back and forth across the paper and from top to bottom, trying to keep the coating as even as possible. Once the paper has been coated, air dry the surface in a horizontal position for several hours, or expedite matters with a hair dryer. Remember, the emulsion becomes light-sensitive as it dries and should be used as soon as possible thereafter. Procedures from emulsion application through development should be done under subdued light (such as a 25-watt yellow light bulb).

A negative transparency is necessary in order to obtain a positive print. Sandwich the negative with the sensitized paper under a sheet of glass. One will quickly find that the paper, with continued wetting and drying, tends to curl. Therefore, optimum print results on multiple color applications can only be obtained with a contact-printing frame of some sort. The frame is required in order to maintain adequate tension between the negative and the paper. Four small C-clamps and a large piece of glass with an identical-sized plywood base will also create the needed tension, sandwiching the negative and the paper in between as in a press.

Expose to an ultraviolet light source: In direct sunlight it should take about five minutes; a sun lamp at 61cm (24in) will need ten minutes. To monitor the exposure place opaque objects on the corners of the covering glass, occasionally removing them, so that you can "eyeball" changes. One will begin to get a feel for exposure times with experience. Generally the greater the apparent contrast between the exposed surface around the objects and the unexposed areas under the objects, the less chance there is for underexposure. Exposure tests by sight are especially useful for bold high contrast images. If one desires a more sophisticated system, a photographic test strip can be made on a piece of identically sensitized paper.

Once the paper has been exposed, the unexposed areas are removed with water. This is the "development" step. Maintain the water temperature at approximately 20°C (68°F). Place the paper, image side up, in a tray of water and agitate gently yet continuously to give the dissolving pigment an opportunity to release itself from the paper. A gentle stroking of the highlight areas with a soft brush during development will also assist the process and give a cleaner image. Change the water three or four times during development. Remember that the emulsion is very soft at this point; so avoid vigorous agitation. Development takes from 7 to 20 minutes.

When the image looks good and all or most of the yellow stain is gone, pull the print and hang it up to dry. Although not always necessary, a four-minute bath in a solution of aluminum potassium sulfate (APS) will assist removal of any remaining orange-brownish dichromate stain. It will also harden the emulsion. It is advised that one use an APS bath between multiple coatings, especially between pigmented layers for three-color separation prints. Keep the APS solution around 10°C (50°F). Prepare a working solution by adding 15 grams of APS to 300ml of water. After the APS bath, rinse the print in a tray of clean water and hang it up to dry.

After the paper has dried, a second color may be added. If the gelatin sizing formula was not used at the start, apply a sparse spray of store-bought starch and repeat the above exposure and development steps, being careful to reposition the negative or place the separation negatives exactly. One can resensitize the print with as many colors as desired.

Realigning negatives on a print for application of additional colors can be tricky. Let's examine a procedure for three-color separation work with same-size negatives. a. Place and tape down negative #1 onto a sheet of pigmented, sensitized paper. Heavily outline the edges of the negative in pen on the paper. Expose #1, develop, and dry the print. b. To align negative #2 for color #2, first register it on a light box with negative #1. Place negative #1 on top of #2, align them so that the images coincide, and carefully slip a piece of double-sided sticky tape between them to keep the images aligned. c. In subdued light, realign the edges of negative #1 (with negative #2 attached underneath it) directly inside the original pen outlines drawn in (a). The print, of course, has been resized and recoated with its second color prior to this step. d. Tape down the edges of negative #2 to the print, wherever the edges of the negative may fall, using tiny pieces of doublesided tape. The tape should not overlap the image area as it might be visually recorded during exposure. e. Carefully remove negative #1 from on top without shifting #2 from position. The paper is now ready to be exposed to negative #2. f. Once color #2 has been processed and the print is dry again, resize and recoat it for color #3. Use negative #1 to make the alignment for negative #3.

This procedure may seem unnecessary and complex, but is very useful should the images not align just perfectly from corner to corner on the transparencies. Regardless of technique, one is still apt to have some problems with registration, as the paper expands and contracts a little with each coating and each development bath.

Below are several helpful hints relevant to gum printing technique:

1. If the emulsion mixture is too thick to put on in an even coat, thin it with water.

2. A thinner coat tends to give better detail.

3. A thicker coating will give a granular effect that works well for bold, graphic images that depend less on fine detail for their impact. It also gives a more three-dimensional effect with multiple layers of color.

4. One can mask the paper selectively to resensitize only a portion of an image. The coating thickness may also be varied with each successive pigmentation.

5. If the image is overexposed, use a hot water bath development. A few drops of household bleach may also help with severe overexposure.

6. For color separation work, try Windsor & Newton Designers' Gouache Colours, printing in the following order: The blue record or separation is printed first in Cadmium Yellow Pale, the green record is printed second in Alizarin Crimson, the red record is printed last in Windsor Blue. Mix a 6mm (¼in) "worm" of color to approximately 15ml gum plus 15ml sensitizer. This will easily coat two 20x25cm (8x10in) prints. These process-like colors are not the only ones that can be used to create a three-color reproduction, and experimenta-

tion is to be encouraged. Also, review the section on how to make color-separation negatives (see chapter 8).

7. Gum prints can be made on a variety of surfaces: wood, plastic, metal, canvas, glass, etc. Gum solutions can also be coated over other images produced by alternative processes, such as cyanotypes or photointaglio prints.

8. Transparency overlays can be applied in the creation of multiple images. One can also combine negatives for each exposure, or an entire image can be altered with each successive coating and new negative combination.

9. If paper is being utilized as the substratum for gum printing, it is also possible to combine recycled image segments from magazines or newspapers with one's own work. (Caution: Be aware of copyright restrictions.) The ink corresponding to the recycled image is lifted off its own base via solvent transfer technique (see below). Once the image transfer is complete, the sensitized gum emulsion may be applied. Photocopier reproductions may also be transferred and combined with gum print images via solvent transfer technique. The photocopier allows the transfer of any image in color or black-and-white, size for size. Images will reverse during transfer. This is usually not a problem as long as there is no written information being transferred.

To transfer an image from a magazine or photocopy, first cover the image with lighter fluid or a mixture of three parts paint thinner to one part xylol (xylene). Position and tape the image face down over the paper to which it will be transferred. Then lightly burnish, or rub, the back of the image with a soft pencil. Different pencil strokes will create varying effects. The solvents loosen the ink, and the action of the pencil completes the transfer. Transfer results will vary from surface to surface. It often helps to blot any excess fluid from the image before taping it down, in order to prevent blotchy results. Sometimes applying the solvents to the blank paper surface rather than to the image surface affords a more even transfer with less image bleeding.

Image transfers can be combined with images produced via other processes besides gum printing technique. The visual impact of a solvent-transferred image can be subtle and worthwhile.

GUM-OIL PRINTING

Oil-based pigment may be employed with gum printing technique to produce images that resemble intaglio prints. This is a direct positive process that employs either a single-weight or document-weight black-and-white print or a positive transparency to produce a positive image. As there is no pigment in the gum, it is possible to expose small images via slide projection; otherwise, use basic contact-printing technique. The greater the image contrast, the greater are chances of success when using a slide projector.

To make a gum-oil print, coat a piece of paper, in subdued light, with a sensitized but unpigmented lithographer's gum arabic solution (14° Baumé). Dry the surface with a hair dryer and expose it.

When the image has been exposed sufficiently, develop and dry it, using heat from a hair dryer to harden the undissolved gum. The

Left and below right: Untitled gum oil prints by John Borna
Below left: Untitled gum oil print made from a paper positive. Original image was made on a high speed, grainy film (Kodak 2475 Recording Film), by Kent E. Wade.

print will appear to have a very faint orange-brownish stain that represents the image in negative form. Once the image is completely dry, lay several "worms" of an oil-based paint (such as Romney Ivory Black) on the surface of the print. Work the paint into the paper, applying more as required. Use standard printmaker's inking technique. The hardened gum will repel the paint, but the paper fibers will be stained by it. Rub the surface with several pieces of cheesecloth until no more ink can be liberated. At this point, either the print can be considered finished or the dichromate stain can be removed with an alum bath. A third option is to remove the gum and dichromate stain completely with bleach: Soak the print in a household bleach bath (150ml of bleach to 600ml of hot water). Do not use a brush to aid removal of the hardened gum or the print highlights might become tainted with color.

Images produced on Kodak 2475 Recording Film will give a very grainy, aquatint like effect. If one desires to print on colored paper surfaces, mix some white pigment into the gum and use an alum rather than a bleach bath after the inking stage, to brighten up the highlights.

THE DUSTING-ON PROCESS

The dusting-on process is a delightful alternative to gum printing. It is a reasonably simple technique that can be mastered quickly. A formula and a procedure for making powder prints on paper or on a glass substratum is offered below.

Part A: 28ml bees' honey
          30ml lithographer's natural gum arabic (14° Baumé)
Part B: 30ml ammonium dichromate
          150ml water.

Add the honey to the gum and mix together in a dish sitting in a tray of hot water (40°C; 120°F) to disperse the honey evenly. (If a thinner mixture of A is desired, use 10ml honey, 15ml gum, and 15ml water.) Next combine the ammonium dichromate and water to make part B. To make a working solution, mix two parts of A to one part of B. There are a multitude of variations to the above formula. Some recipes call for sugar instead of honey. Others utilize fresh egg albumin, LePage's fish glue, gelatin, etc., as the colloid to replace the gum arabic.

Coat the paper or printing surface with the working solution. Use a soft, 5cm (2in) wide brush. Dry the surface of the paper quickly with a hair dryer, being careful not to bake the emulsion's surface by overheating so that it loses its stickiness. A second coat is optional and is brushed on when the first coat is dry. Coating and drying steps should be done in subdued light.

This is a contact-speed direct-positive process, therefore a positive transparency is needed. Place the transparency over the slightly tacky surface and cover it with a sheet of glass. High-contrast positives work best. Next, expose this sandwich to an ultraviolet light source: a sun lamp at 60cm (24in) takes approximately six to ten minutes. Do not place the light source so close to the emulsion as to bake it.

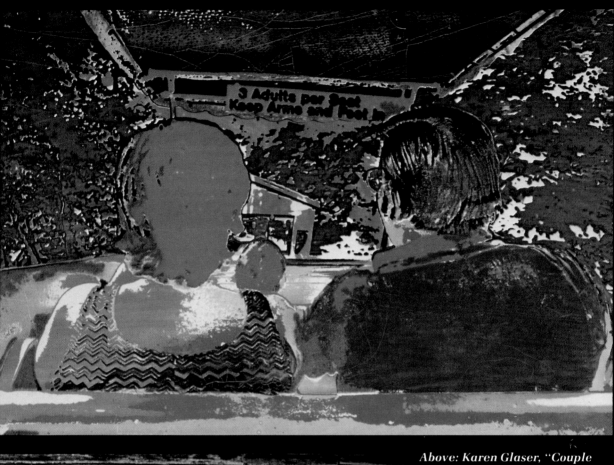

Above: Karen Glaser, "Couple Riding an Incline." A gum print on B & W resin-coated paper. Eight colors were used, coating with several colors at one time prior to each exposure (approximately 5 total). The black in the print represents the developed silver image. 4" x 6". Left: Kent E. Wade, "Bicentennial Portrait of America—1976." Handcolored B & W photo using Peerless Watercolors.

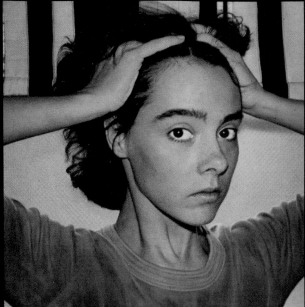

Above: Gail Skoff, "Tinda." A B & W print, selenium toned and hand colored with Marshall's photo oils and pencils 12" x 12"
Above right: Colleen Kenyon, "Colleen, Fall 1976." Handcolored print on Portriga Rapid, semi-matte paper with additional sepia toning. Marshall's photo oils and pencils were used. 14-5/8" x 14-1/2"
Right: Christopher P. James, "Wicker Pattern." B & W print, toned sepia via a sodium sulfide toning process. Flat colors created by use of rubber cement masking/toning technique. A textured quality was created by tinting dried surface with opaque and transparent enamel pigment, scribing the unhardened enamel, pressing texture into it, buffing it, etc.
Below Right: Ford Gilbreath. Untitled. Handcolored B & W print using Marshall's photo oils on Kodak Poly G (matte surface) paper 3" x 5"

Facing page: Robert Embry. Untitled. Balcony image is from a Mexican portfolio of dye transfer prints. Color separations were made from an Agfachrome 50-S transparency. 10-3/4" x 8-5/8'

Facing page, top: Stanley R. Smith, "False Helibore." A dye transfer print.
Below: Steven J. Cromwell, untitled. D-12 dye transfer print from 3 matrices.

This page top: Judy Durick, "Showing the Scars." Photo-lithography printed in 5 colors from 5 separate negatives and aluminum plates. 19" x 18-1/2"
Bottom left: Kathleen Kenyon, "Heavenly Bodies." A photo-lithograph. 15" x 20"
Bottom right: Paul Miller, "Tree." A photo intaglio print of 3 colors, handwiped and rolled-on. 12" x 16"

Facing page, upper left: Bea Nettles, "Moonrise through the Pines." A Kwik Print image. 20" x 26"
Upper right: Larry Bullis, Scotty Sapiro. Untitled. Screen print, 3 transparencies were color separated (red, green and blue filters) from a slide onto pan film and positive transparencies of varying densities were made from these separations. 5 colors were used: a yellow, blue and red transparent as well as a black and gray opaque ink. 9-1/4" x 13-3/4"
Below: Colleen Kenyon, "The First Parting Picture." A screen print on Rives BFK paper involving 15 separate colors. The original image was a

B & W negative, tonally separated into 10 different densities. These transparencies were processed for graininess. Both positive and negative separations were employed for printing. 15-1/2" x 15-1/2"

This page, above: Fredrich Cantor, "Aura." A 3-color gum bichromate print. The negative separations were made from a 4-color litho proof of a manipulated silver print. 4-3/4" x 7"
Left: Richard A. Kyle, "Gears." A gum print on Arches Cover using 7 separate exposures and coats of watercolor pigment.

Above left: Ellen Land Weber, untitled. 3-M Color-in-Color image transfers. Copier images were made from a leaf and a picture. Both were reproduced separately on the machine and transferred to Arches heavyweight paper, initially sized with 3-M Color-in-Color Sizing Concentrate. The transfer was made with heat and pressure from a drymount press.

Above: Jill Lynne, "Triptych #3: Madam Lorraine and the Peacock Feathers." 8-1/2″ x 14″

Left: Barbara Astman, "The Sun Ritual." A Xerox print. The image was made by reproducing a hand-tinted B & W photo which was sandwiched with an acetate overlay. The swans were cut out and applied to the acetate screen. 8″ x 10″

Development is a unique procedure: Very gently expell a gush of hot, humid air from your mouth onto the surface area to be dusted. Humid moisture tends to make the unexposed surface a bit tackier. A drinking straw will also prove extremely useful here, as it affords selective control in developing the image.

Almost any very fine powdered pigment (400-to-600 mesh) can be used to develop the image. Apply the powder by dusting it on with the help of an enameling sieve or similar device. Try a small bottle, its opening covered with a nylon stocking. As the powder is applied, alternate between pushing it about the unexposed areas with a fine brush and breathing humid air on the same area. This does take some practice; so proceed patiently. Do not breathe too hot-and-heavy; this tends to make the highlights (exposed areas) slightly tacky again. Generally, if an image comes up too dark it is probably underexposed. If it is too light, it is overexposed. Once the image has reached a satisfactory depth of color, expose the image again to a UV light source until all remaining tackiness is eliminated. A thin coating of rubber cement or decal covercoat will protect the image.

The powder process can also be combined with transfer technique for decorating objects in decal fashion. Here a sheet of glass is employed as a temporary support. The image is first developed with powder as described above. Next it is coated with collodion. The image is then rinsed in an alum solution to remove any dichromate stain. Finally the powdered image is carefully separated from the glass plate and affixed to the selected surface. In this manner finely ground enamel- or ceramic-based images can be transferred to copper or porcelain and fired on for permanency. Modifications of this technique also enable acid-resistant images of finely powdered bitumen to be transferred from paper to glass for the purpose of etching the glass.

SUN-BLEACHED IMAGERY

During the 1830s and 1840s, Robert W. Hunt experimented in the use of natural plant pigment as a light-sensitive compound to produce photographic imagery. The results proved to be more amusing than useful, as images tended to self-destruct upon re-exposure to any sunlight. For those who are curious and would like to experiment with this process, the following information is offered.

Anthotype, or nature printing, is very simple and involves extracting dyes from freshly picked flowers, sectioning and pulverizing the plant sections, petals, berries, etc., with alcohol, then coating a piece of paper for a sun exposure. The exposure can take hours, weeks, or even several months, depending upon the dye matter. The sun bleaches, or fades, the colored tincture, thus a positive transparency is required to produce a positive picture.

Experiment with dyes from vegetables, as well as from flowers. Images have been produced using beet and carrot tinctures. Red-pigmented and blue-pigmented tinctures produce stronger images with better contrast. Mordanting technique (do some research on fabric mordant and natural dyestuffs) may help to improve the fade-resistance of the final print.

Mounting Technique

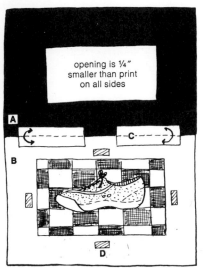

opening is ¼"
smaller than print
on all sides

A

B

C

D

A & B are identically sized pieces of matte board
C: packing tape hinges   D: 2 sided sticky tape

Lastly, bleached imagery is also possible using different colors of construction paper. Experiment with the darker colors for best results. Photograms are easily attained by laying objects on the paper and placing them in direct sunlight until the exposed color has faded out.

MOUNTING TECHNIQUE FOR PRINTS ON PAPER

To mount a paper print, including gum bichromates, cyanotypes, photosilkscreen, or photointaglio prints, first position the image on a piece of good-quality mat board with double-sided sticky tape, leaving equal margins around top and sides of the print. Many people prefer to have half again as much margin at the bottom. Take another piece of mat board the same size and cut a window in the center. To do this, carefully measure the exact width of the margins on the first board (the mount) from the edge of the print out to the edge of the mount. Mark that distance on the wrong side of the new board (the mat), then draw an outline at least 6mm (¼in) inside the marks and cut on this line. The window opening, being smaller than the print, will thus hide any untidy print edges. A good mat cutter (Logan Hand Mat Cutter #301, X-Acto #110, or Dexter mat cutter) will save a lot of grief in making this opening. The best substitute, if you do not have a cutter available, is a one-sided razor, used in conjunction with a metal ruler, a masonite-topped table, and a steady hand. Next, lay the windowed mat over the print on the mount and align the edges of the boards. Keeping the tops together, raise the mat from the bottom, as though opening a book. Lay it face down on the table with its top against the top of the mount board. Run two or three strips of an acid-free linen tape (library type) or packaging tape from one board to the other to act as hinges. Fold the mat back over the mount and even up the edges again. If necessary, reposition the image, then insert a piece of double-sided tape at the bottom to hold the two boards together. The mounted print can now be placed in a frame. (Prints on textile surfaces may also be mounted using this technique. It is suggested, however, that one leave the glass out of the frame so that the texture may be observed more clearly.) Lastly, where a frame is not to be employed, place more double-sided tape under the corners of the print to secure it and also around the window opening to prevent gapping.

The Photographic Transparency

Most of the photographic printing processes discussed in this book utilize a contact-exposure system, and thus an enlarged black-and-white transparency of the original negative, print, or artwork is usually needed. These enlarged transparencies can be made either with conventional continuous-tone sheet film or with high-contrast lithographic film. High-contrast films are designed to drop out middle gray tones in order to produce strong blacks and whites. The creative manipulation of these films offers a wide variety of additional image possibilities. Some of these variations will be presented here, along with a basic approach for producing high-contrast transparencies, both positive and negative.

HOW TO MAKE HIGH-CONTRAST POSITIVES AND NEGATIVES

For years graphic artists have been using black-and-white films that are high in contrast, insensitive to red light (orthochromatic), durable, reasonably slow, and packaged in a variety of sheet sizes. Several companies manufacture these high-contrast films. All are processed and handled in a similar manner, not unlike photographic papers. One reliable brand is Kodak Kodalith Ortho Film 6556, Type 3. It is an excellent film for making high-contrast negatives or positives. Another brand is 3M Line Ortho PL-5 film. It is an inexpensive film well suited to high-contrast line reproduction. Both products can be processed in Kodalith A & B Developer.

Enlarged negative or positive transparencies can be prepared in any basic darkroom facility or purchased commercially. Special effects, including line images, halftones, texture-screened transparencies, and tonal separations, can also be obtained at reasonable cost. Consult the larger print shops, blueprint firms, or photo labs.

The procedure for making an enlarged high-contrast positive from a camera-size negative is outlined below.

1. Place the negative in an enlarger and project the image onto a sheet of plain white paper in order to focus and to determine the size

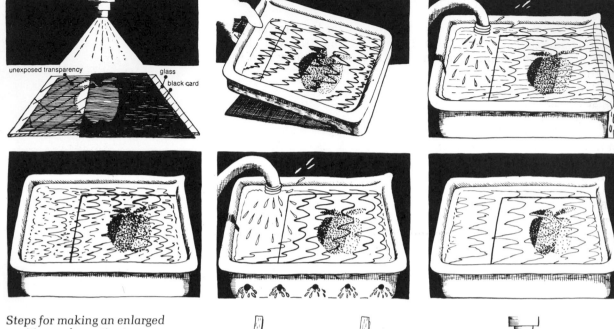

Steps for making an enlarged positive and negative transparency: (left to right from top).
1. Expose the high contrast film emulsion side up (dull side) to a projected negative. Run a test strip in 3 second increments using a black opaque card. *2.* Develop the film for 2½ minutes. *3.* Water rinse, 45 seconds. *4.* Fix for 4 minutes or until film base clears. For final transparency choose best exposure time and repeat steps 1 through 4, then proceed with 5 through 7.
5. Wash for 12 minutes. *6.* Kodak Photo Flo-200, 1 minute with agitation. *7.* Hang to dry. Save test strips as indicators of other possible tone effects for further manipulation of an image. *8.* To make an enlarged high contrast negative—sandwich the enlarged positive with a sheet of unexposed film (emulsion to emulsion) and expose. Process the same as steps 2 through 7.

of the enlargement. Next, under red safelight, replace the paper with a sheet of high-contrast film, dull (emulsion) side up. Use a piece of clean single-strength (thin) glass to hold the unexposed film flat.

2. Set the lens aperture at f/11. Run a test strip in three-second intervals, using normal photographic procedure; that is: Cover all but one strip of the unexposed film on the enlarging easel with a piece of opaque cardstock and turn on enlarger. Every three seconds uncover a similar width of film until the entire sheet has been exposed to the light. If five separate three-second exposures are made, then the first strip exposed will represent 15 seconds of exposure; the second, 12 seconds; and so on.

3. Develop the test strip in Kodalith A & B Developer. Both parts can be premixed and stored in separate containers until needed. Just prior to a working session, mix together equal amounts of part A and part B. Develop the film for 2½ minutes, with continuous agitation, at 20°C (68°F).

Sometimes it may be useful to hold more grain or tone in the transparency without having to purchase another type of sheet film. This can be accomplished by diluting Kodak Dektol 1:12, using it as a substitute film developer. The remaining processing steps will be the same.

After development, rinse the film in a water bath for about 30 seconds and then immerse it in fixer, where the film will clear. It takes about three minutes for the film to fix completely in a fresh solution.

Remove the test film and examine it under white light. Choose the strip that had the shortest exposure time affording a deep black shadow area; that is, the first strip beyond which additional exposure increments no longer appear to make an appreciable difference in the "blackness" of the tone. You have now established a normal exposure time for this negative.

4. To make the positive transparency, repeat steps 1. and 3., using the newly established exposure time.

5. After the film has fixed, turn on the light. If any midtones remain that are objectionable, these are now removed either by contact-printing the transparency onto another piece of high-contrast film or by using a chemical reducer (see section on reducing silver bromide prints). By eliminating these transitional areas of weak density, strong graphic images are possible. Sometimes, during chemical reduction the black areas of the transparency are also reduced considerably. In these instances, density and contrast can be increased again by chemical intensification of the image. For chemical reduction and intensification, try Farmer's Reducer (potassium ferricyanide) and chromium intensifier. Both are put up by Kodak in small packets and are available at most photographic stores. Mix and use each according to directions packaged with it.

6. Once the positive transparency has been treated, wash it for about 15 minutes in running water.

7. Finally, bathe the film in a solution of a wetting agent such as Kodak Photo-Flo 200 for about 40 seconds with continuous agitation. A wetting agent helps to prevent watermarks while the film dries. Hang the film by one corner. Always hang tonal separations from the identical corner in relation to the image. A hair dryer can be used to expedite drying time.

In order to make an enlarged high-contrast negative, place the enlarged positive transparency on top of a fresh sheet of litho film the same size, resting on the enlarging easel or baseboard. The two pieces of film are held sandwiched together, emulsion to emulsion, with a piece of clean glass. Be sure the negative carrier in the enlarger is empty. Again, make a test strip to establish the best exposure time, using an aperture of f/11. Maintain a distance between the light source and the unexposed film equal to the approximate length of the film's diagonal, in order to insure even coverage.

An enlarged high-contrast negative can also be made by the litho etch/reversal technique (review Chapter 6). In addition, if one is working from a color slide a large negative can be made directly by unmounting the slide and enlarging it onto a sheet of high-contrast ortho film. Remember, however, that orthochromatic films are blind to red and may give a false rendition of the subject. In instances where a slide contains a great deal of red, or where a high-contrast color separation is desired, use a high-contrast panchromatic film, such as Kodalith Pan Film 2568. This film must be processed in total darkness.

Occasionally a high-contrast negative is desired for a process permitting exposure by slide projection. In this case, copy the original work with Kodak High-Contrast Copy Film and slide-mount the negative. If one desires graininess in the negative, then try high-speed black-and-white Kodak 2475 Recording Film for rephotographing the work. For those images already in color-slide form, contact-print the slide onto a high-contrast panchromatic or orthochromatic film and slide-mount the resultant negative for slide projection.

Enlarged negative transparencies are sometimes made by contact-printing enlarged black-and-white paper positives onto litho film. Any single-weight or document-weight projection (enlarging) paper without writing on the back will suffice. Thicker, translucent materials are not recommended, as exposure times become excessive. Finally, "positive-reading" drawings on acetate or thin tracing paper can also be reversed by contact-printing. If necessary, tracing paper can be made more transparent by waxing or oiling it.

RETOUCHING THE TRANSPARENCY
All litho-type films have a tendency to develop pinholes. These can be blocked with any black or red opaquing liquid purchased from a graphic arts supplier. Try Perfex Black Opaque. It is premixed to a fine working consistency. It will also wash off with water, should mistakes be made.

One is not limited to just touching up pinholes. It is possible to eliminate virtually any undesirable element or area in a transparency by painting it out with opaque. Thus, opaque is another creative tool in image construction. Use a high-quality sable-hair brush, #00, for pinholes and fine detail work and a larger brush for painting out unattractive backgrounds or subject matter.

Lastly, when it is feasible do as much touch-up work as possible on the first transparency before contact-printing it to a second piece of film.

SPECIAL IMAGE EFFECTS USING HIGH-CONTRAST FILM
There are many ways to produce special effects on high-contrast film for image-making. Many of these effects can also be combined, increasing the number of image variations that are possible.

1. Tone Separation. This is a technique for separating the density range of a continuous-tone image into high-contrast "building blocks" for creative image reconstruction. Separation is done through exposures: the midtones through normal exposure, the shadow areas through underexposure, and the highlights through overexposure. (The procedure will be explained under a. below). A positive set of these tonal segments is made by contact-printing a negative set, and vice versa. These transparencies are processed like any other image on high-contrast film. Image reconstruction may take several forms, including posterized images, line images, and images resulting from the sandwiching of positives and negatives of different densities. Tone separations can be used to reproduce graphic images on almost any surface material in conjunction with a variety of photographic processes.

a. Posterization. By separating a continuous-tone image into a *minimum* of three positives or negatives of different densities, one can reassemble the image into a semitone, or posterized, reproduction.

Make a set of high-contrast positives by exposing one for normal density (per test-strip results), another at half that exposure, and a third at twice the normal exposure. The normal exposure will give the midtones of the image, the half-exposure will contain shadow information, and the twice-normal exposure will be used to print the highlights. If required, these transparencies can each be contact-printed to make a *set* of enlarged tonally separated negatives.

To create a posterlike image in color, the density, or tone, separations from the set are used individually to print several superimposed layers of pigmented emulsion, one at a time. A posterized gum print can be made this way, using a set of negative transparencies, plus, perhaps, a highlight positive for printing the background in.

The technique is different for making posterized images using silkscreen. For this, the density separations from a positive set are used to make photostencils. Each stencil is then used to screen one layer or color of ink.

b. Line Images. Line effects that look like pen-and-ink drawings can be produced photographically from separation transparencies. A line image is formed by sandwiching the positive and negative high-contrast separations of the same density of the same original image. For convenience, a line mask is usually made on a separate sheet of litho film by contact-printing. The line mask should be made with the exposing light source at an *angle* to the film plane. Further image variations are possible by printing with several line masks from the same image.

2. "Solarized" Images. Unusual images that appear to be both positive and negative at the same time can be made on high-contrast film by momentarily re-exposing the film to light during development. This phenomena is known as the Sabattier effect. To "solarize" a litho film, follow this procedure:

Expose the film only about 60% of its normal exposure time. Develop, preferably using a brown-looking, or "overworked," developer, as this tends to produce more pleasing solarized images, with more luminous whites. When the film is 75% developed, remove it from the developer, lay it on a clean tray or other flat support, and wipe it with a sponge. Place the partly developed transparency under the enlarger. Set the enlarger lens at f/16 and double check to be sure there is no negative in the enlarger's carrier. Run a test strip in three-second intervals and complete development; however, this time do not agitate, except for a few seconds initially to make sure the film is completely covered with developer. Rinse and fix the film as usual. Having chosen the strip you prefer, make a final transparency following the same steps and using the selected re-exposure time. An optional re-exposure technique is to let the film settle to the bottom of the developer tray, then expose it briefly to a 25-watt bulb about one meter away.

Many factors will affect the outcome of solarization: the length of the first and second exposures, the age and dilution of the developer, the length of first and second development, and so one. It is therefore a good idea to maintain accurate records in order that pleasing results can be duplicated successfully. As a final point, density separations can also be solarized, to be integrated later with other separations during printing.

3. Multiple Images. An image can be repeated by printing it several times in different locations on one transparency. To do this, make a mask of black paper to keep parts of the litho material unexposed while exposing other sections. Be sure to cover up parts that have been exposed previously, as well. This technique may be used to add more than one image to a sheet of film.

4. Image Collages. Images can be cut out of litho film and several assembled together onto a sheet of Mylar or acetate to form a collage.

5. Texture Screens. A textured design or pattern can be combined with any high-contrast image for added visual impact. This is accomplished by the use of texture screens. The texture can be supplied by a piece of frosted plastic, a greasy substance on glass, a piece of fabric. The screen is made by photographing the pattern or contact-printing the texture. The pattern is incorporated with the image by projecting the image through the screen, sometimes through the material itself, as it lies on top of the unexposed film. If the pattern was recorded by camera, then it can be used in its smaller negative form by sandwiching it with a negative of an image to be enlarged. This composite can then be projected for printing.

Contact screens for special effects can also be obtained from suppliers of graphic arts materials. There are many patterns available, identified, for example, as Steel Engraving, Mezzotint, Linen, Copperplate, and so on. In addition, small texture negatives for projection can now be purchased at photographic stores.

Another graphic arts tool useful for adding texture to an image is dry-transfer material. Here, the pattern is integrated with the image by burnishing, or rubbing, the design off its temporary support. This material is useful for the selective application of texture, reducing the need for intricate masking during exposure (see Chapter 10).

6. Transparency as Print. The positive transparency itself can be used as a final form of photo imagery rather than as a tool for image reproduction. For example, several positive transparencies that could be used for creating a posterized print might also be hand-colored, toned, or dyed, then aligned and mounted in a viewing box. Images thus assembled would produce an effect similar to that of color-proofing films (see Chapter 10).

HOW TO MAKE A HALFTONE TRANSPARENCY ON LITHO FILM

A halftone transparency is one in which the image is recorded as small opaque dots that vary in physical size and frequency but that suggest the appearance of continuous tone. If one requires only a few halftone transparencies, these can be made by utilizing a special film called Kodalith Autoscreen Ortho Film 2563 (Estar Base). This film has a dot pattern built into it and is processed the same as other litho films. In order to differentiate the various finenesses of halftone screens available, one refers to the dot pattern by number of dots per square inch, arranged in lines. Autoscreen film is rated at 133 lines, which reproduces as a relatively fine pattern (dots being barely discernible to the naked eye). Unfortunately, a 133-line screen is too fine for photosilkscreen application, as the mesh has difficulty holding that fine a dot pattern. In order to obtain a coarser screen (e.g., 65-line to 40-line), either purchase a commercially made halftone transparency or experiment with enlarging a smaller negative made on Autoscreen material. Enlargement will make the dots and pattern larger, closer to the desired 65-line size.

An alternative to Kodalith Autoscreen Ortho Film is the halftone screen for contact-printing applications. These are available in different screen sizes and line ratings. Kodak's 20x25cm (9x11in.) 65-line elliptical-dot gray contact screen (negative) is an ideal screen for smaller work and suitable for use with all processes. An image that has been produced with a 65-line screen will exhibit a visible dot pattern, but this is not necessarily unpleasing. If one desires a little finer screen, try an 85-line (typical of most newspaper work). The 65-line, however, works best for photoetching, enabling one to etch a little deeper than would be possible with a finer screen. With a tighter line pattern the dots are easily undercut and the image highlights and upper midtones rapidly disappear.

To use a halftone screen, simply sandwich it with a sheet of Kodalith Ortho Film 6556 under glass and expose it to a negative via an enlarger. If a negative halftone is desired, contact-print the positive halftone to a fresh piece of film, emulsion to emulsion.

When making halftones, bear in mind the importance of having some dot pattern visible in both highlight and shadow areas. Expose the film initially to pick up a dot pattern in the highlights. Then remove the negative from the englarger and flash the screen/transparency sandwich for a few seconds. A flash test strip will prove useful. Flashing ensures a longer tonal range from white to black by breaking up the shadow areas with a dot pattern.

ALTERNATIVE TRANSPARENCY FABRICATION METHODS
It isn't really necessary to leave one's studio in order to generate new imagery. Some possibilities involve the manipulation and/or collaging of several images already on file. Often images or image segments found in magazines, newspapers, etc., can be successfully combined with one's own artwork, although of course one must be aware of copyright restrictions when using images in this manner.

An image can be lifted out of a clay coated magazine by coating the picture surface with Liquitex Gloss Polymer Medium or a similar product available from any artists' supply store. Brush on about five coats, drying in between. Next, immerse the coated print in warm water, slowly working off the paper backing. What remains is a transparency in full color and likeness to the original recycled image, on a flexible skin of medium. The ink from the magazine has been absorbed into the medium. Denatured alcohol will quickly remove any unwanted portions of the image. Images thus prepared can be distorted by stretching or be used unaltered as a positive transparency. To make a negative transparency, contact-print the positive image onto a sheet of high-contrast ortho or pan film.

Another method for transferring images from a newspaper is to coat a sheet of transparent acetate with rubber cement. Once it is dry, place the newspaper image face down on top and burnish with a soft pencil. The ink is drawn out of the paper into the rubber cement by the rubbing action. Any touching up of the image with pencil prior to burnishing will transfer, as well. The image on the acetate can now be employed as a positive transparency.

Sometimes hand-drawn work can be added as an extension of or as a complement to a photographic image. A transparency can easily be made by drawing on clear acetate or thin tracing paper with India ink or a photographic opaquing solution. One can also add a drawing to a processed lithographic transparency.

A transparency is not necessarily required to produce an image. One can make photograms directly at the exposure step of the particular process involved. Rather than producing a high-contrast intermediate, place objects (flowers, hands, etc.) directly on the light-sensitive emulsion. Exposure to light will cause those areas hidden under the object to remain unexposed, resulting in a silhouette. Not only can this be a great deal of fun and highly creative, but it is also a simple way to involve children in some of the less complicated photographic processes.

BASIC COLOR SEPARATION TECHNIQUE

Color separation is a step in the reproduction of colored images in like colors. Highly exacting color reproduction requires refined techniques and special equipment, such a densitometers, constant-voltage regulators, special film masks, etc. It is possible, however, to obtain reasonably good results without these devices. The procedure that follows, which comprises several contemporary modifications, can be used to reproduce successful color images comparable to those achieved in the infancy of color printing. In addition, any such experimentation should lead to a better understanding of the rudiments of color theory.

One of the ways in which one may utilize color-separation technique is in the three-color reproduction of images by the gum bichromate, photosilkscreen, photointaglio, or photo-offset processes. Another possibility is to use separation negatives to produce three-color dye-transfer prints. The colors provided by dye transfer are indeed beautiful. Images appear to have a three-dimensional quality and depth unmatched by most color-print photographic papers. Three color-separated negatives, made by filtration of the primary photographic colors, are contact-printed or projection-printed onto three gelatin-coated matrices of Kodak Matrix Film 4150. Each of the three matrix images is dyed in a solution of the complementary color of the filter used to make its corresponding negative. A full-color image is reconstructed on Kodak Dye Transfer Paper or Film by aligning, or registering, the three matrices in turn, transferring the dye from each one in succession. Multiple images can be produced from one set of matrices. It is also possible to

Color separation aids

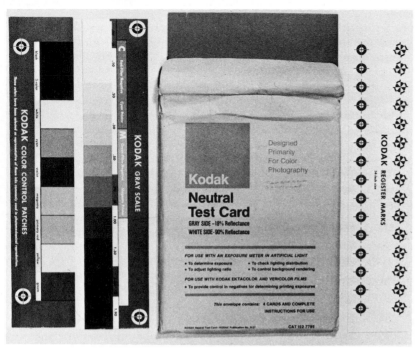

prepare alternative surfaces with several coats of a gelatin-based or liquid photoemulsion, mordanting the gelatin for transfer of the dye. Nontextured surfaces will afford a better transfer. Bleeding of the image is always a concern, and experimentation will be necessary (review the dye imbibition process for matrix preparation and dye transfer technique).

HOW COLORS ARE SEPARATED

In order to reproduce a subject in color, one must first separate the colors on black-and-white continuous-tone film. Panchromatic film is required, as it is sensitive to all the colors of the spectrum, whereas orthochromatic film is blind to red. Three negatives are to be made by utilizing the different primary filters: red, green, and blue. All the colors of the spectrum will be represented when these three negatives are printed in register, in its complementary color: the complement of red is cyan (blue-green); the complement of green is magenta (blue-red); the complement of blue is yellow (red-green).

It is very important that the three separation negatives be of similar contrast and density. In this way, one has some assurance that the colors of the original subject will be reproduced in balanced proportions. Such balance is best judged by placing a gray scale in the scene when making the color separations. In theory, all three filters should produce an identical step wedge, or scale of gray tones; that is, each of the three negatives should exhibit similar density in the white, in the gray, and in the black areas. Examination of the gray scales, best reviewed in prints made from each negative, reveals certain useful information: For example, overexposure of a negative is apparent if all the increments of its gray scale are each darker than all the steps in another separation. If the lighter gray steps in two negatives appear similar but one of the scales goes to darker grays sooner, then the latter negative, with its compressed tonal scale, is too contrasty and was overdeveloped. Remember, too, that color balance will be more critical in the highlight areas of the print. Thus, the denser portions of the negative (the light steps on the wedge in the prints) should be reasonably close on all three.

It is difficult to tell, on black-and-white film, which color-separation negative was produced with what filter, although the green-filtered negative will appear to be the most like a normal black-and-white negative. To avoid confusion, add color patches along the edge of the subject when making exposures. Each filter passes its own color, while absorbing the others, and thus each filter's own color will be represented by the densest, or blackest, patch in each negative, or as the lightest patch on a print. For greater visual ease, make the color patches in letter shapes relating to the color filters being employed. One can also code transparencies by notching the corners of sheet film prior to exposure: red filter=one notch, green filter=two notches, blue filter=three notches.

Once the scene or artwork has been recorded on three separate negatives, the films must be processed. The blue-filtered negative should receive about 30% to 45% more development than those filtered with red or green. Film emulsions are inherently sensitive to blue light. A blue filter increases the response of the film to blue, thus lowering the contrast of the negative. Increased development will help to increase the contrast of the negative. If one wants to be exacting, the red-filtered negative should receive about 15% less development, to reduce contrast. The gray scale, of course, will indicate even more accurate development times. Establish a development time for the green-filtered negative and use it as a norm from which to vary the other development times.

If one intends to make separation negatives on 35mm roll film, then it becomes rather difficult to dicuss custom development times. Actually, one will find that development of all three negatives to a normal development time will not give as fine a separation but will give reasonably good results in the final color reproduction. Remember to dry all separation negatives, especially those on sheet film, hanging in the same direction in reference to the image.

A fourth separation negative can also be made. It is a contrasty black separation negative designed to add depth or density, but only to the shadow areas. This fourth negative can be made by exposing a sheet of film through all three filters, giving one-third of the exposure time to each. As an alternative, one might also try using a Wratten K2-filtered negative as the black separation. Overexposure and overdevelopment of this fourth negative will help to achieve the desired result.

HOW TO MAKE COLOR SEPARATIONS
FROM VARIOUS SUBJECTS

There are several sources from which one may choose a colored image for reproduction. Following is a discussion of these sources and of the different approaches to making color separations from them. In all cases, the inclusion of registration marks and filter-color signals on the edges of the subject area during exposure can be helpful later in re-aligning and identifying separations.

From Life. Color separations can be made from original subject matter. It is possible to expose a scene with the aid of a stable, solid tripod using three filters: a red (Wratten #25), a green (Wratten #56), and a blue (Wratten #47). It will be necessary during exposure to compensate for the added density of the filters. Because all three filters have a similar filter factor, either open up the lens three extra stops from the normal reading or adjust the shutter speed accordingly. Generally, it is preferable to make the adjustment on the shutter-speed dial in order to maintain maximum depth of field. To make the exposures, one can use a 35mm camera and Kodak Plus-X or equivalent film, or any larger-format camera. In the latter case, and when using sheet film, try Kodak's Super-XX Pan Film 4142. Develop Super-XX Pan in HC 110 according to film instructions for separation work. Finally, all exposures and filter changes should be made without any movement of the camera.

If silkscreen positive separations are required, the three separation negatives may be enlarged at this time. If larger negative separations are desired, as would be necessary for a gum print, one can enlarge the smaller separation negatives onto a 20x25cm (8x10in.) sheet of Kodak Professional Direct Duplicating Film SO-015. This is an ortho film that produces a continuous-tone negative from a negative. An ortho film can be utilized once the original color separations have been made because the gray tonal scale will be recorded accurately even under red safelight.

There are also other options available for making enlarged negatives. For example, one can contact-print the original separation negatives onto Kodak Commercial Film 6127 (a moderately high contrast continuous-tone ortho film) and then enlarge these positives onto sheets of high-contrast litho film. By developing these enlarged negatives in one part Kodak Dektol developer to twelve parts water, a grainy continuous-tone transparency suitable for gum printing is obtained. Develop the film for approximately 1½ minutes. Add more water if the developing time is too fast.

One can also employ 35mm Kokak Direct Postive Panchromatic Film 5246 for making black-and-white continuous-tone positives of the original subject. These smaller positive separations can then be enlarged onto high-contrast litho film and developed in diluted Dektol for an increase in tonality, or enlarged onto Kodak Commercial Film 6127 for optimum continuous-tone results.

From Artwork. Color separations can be made from color prints. In making such separations, the same procedures as outlined above would apply. Here, the critical factor is proper lighting of the print while it is being rephotographed.

Place or suspend the print so that the surface of the artwork is parallel to the film plane. Place a 500-watt photoflood on each side of the camera at a 45° angle to the print-camera axis, each exactly the same distance away from the print. It is very important that the print be evenly illuminated. A gray card is a useful tool to assist in establishing an accurate meter reading. Take the gray-card reading along the edges, at the corners, and in the middle of the print. If readings are not the same, readjust the lights. A slightly higher, yet consistent, reading around the edges is acceptable. Once the lights are properly adjusted, make the three filter-separated negatives.

From Color Transparencies or Slides. Color separations can be made from positive color transparencies. A color-correction mask is useful, as it can lower the contrast range and alter the color saturation of the transparency. The mask is made by contact-printing the transparency onto Kodak Pan Masking Film 4570, in sandwich with a diffusion sheet to make the mask slightly unsharp and easier to register. Exposure is made via a white or colored light source, depending on the color correction desired. A correctly processed mask will appear low in contrast but exhibit detail in both shadow and highlight areas. Once it is prepared, the mask is then bound with the transparency for the filter separation step. If maximum color reproduction is desired, a second mask can also be made. Remember, however, these masks are a refinement only intended to improve color balance. They are not a "necessary" for experimental work.

Color separation negatives can be made by contact printing, enlarging, or rephotographing the color transparency. If one is using an enlarger, it is advisable to rig up a jig, so that sheets of pan film or a sheet film holder can be handled easily in total darkness. If one is copying a transparency, use a slide copier. To make a homemade slide duplicator, cut a hole in the center of a large sheet of nonreflective black cardstock; place the transparency over the hole and tape it down; pin this card in a doorway. Illuminate the transparency from behind, using a 500-watt photoflood lamp diffused so that the lighting is even. Using the appropriate film, camera filters, extension tubes, and tripod, make exposures from the front side. The only light during exposure visible to the camera's lens should be that passing through the illuminated transparency. Use the following filters when copying a transparency: Wratten #29 (dark red), a Wratten #58 or #61 (dark green), and a Wratten #C-47 or #47B (dark blue).

From Color Negatives. To make color separations from color negatives, first make three positive prints, one through each of the appropriate filters, using Kodak Panalure paper. Pan paper is designed for printing color negatives in continuous black-and-white. Although this paper is panchromatic, it can be used with a low-wattage dark amber safelight; otherwise it should be used in total darkness. After processing these positive separation prints, rephotograph them. Use Kodak Professional Copy Film 4125. This film is designed for copying continuous-tone black-and-white prints, and affords optimum highlight reproductions, when compared with other films. You will now have the correct color-separated negatives necessary for three-color image reconstruction.

HALFTONE COLOR SEPARATIONS
FOR PHOTOSILKSCREENING

Halftone positive separation transparencies are required for photosilkscreening a continuous-tone color image. Texture screens, such as a 50% fine mezzotint pattern, may also be used, and the same procedure is advised without the need to angle separations. This section, however, will discuss the use of halftone screens.

Initially, three color-separated negatives must be made. In order to make enlarged halftone positive separation transparencies, either have them made commerically from three 20x25cm (8x10in.) black-and-white positive separation prints or make them in one's own darkroom.

To make a screened positive separation, a halftone screen larger than the size of the image desired will be necessary. For example, a 23x28cm (9x11in.) screen will only cover about a 13x18cm (5x7in.) area when it is angled. Lay the screen down directly on top of a sheet of unexposed litho film on the enlarger's baseboard. Project the first negative and process in Kodalith A & B Developer. Follow this same procedure until all three separations have been made. (Review first part of this chapter, on litho films.)

But before each exposure, however, note the position of the halftone screen and bear in mind that the dot pattern of each separation positive should be at an approximate 15°-to-45° angle to the dot pattern of the other positives. This is done to prevent a moiré effect during printing. (To see what this means, take two halftone transparencies and place them over a lightbox, one on top of the other. As the top one is rotated over the other, a moiré pattern will become visible.) The angle at which this effect is less apparent is approximately 30°. If halftone separations are to be made commercially, then advise the printer how the transparencies will be used, so that the print separations will not all be screened in the same direction.

Below is a procedure for angling the halftone screen properly. Lay a sheet of unexposed 20x25cm (8x10in.) Kodalith film on the enlarger's baseboard. • To make the first separation (the one used to print yellow), place the 23x28cm (9x11in.) halftone screen over the film so that its short side lies parallel to the short (20cm) side of the film and expose, using the blue-filtered negative. • The second exposure (the cyan-printer) is made with a fresh piece of unexposed film lying in the same position as the first, but this time the halftone screen is slanted about 30° *clockwise* when it is laid down on top. Expose, using the red-filtered negative. • In the third exposure (the separation for printing magenta), the screen is slanted approximately 30° *counter*clockwise while printing with the green-filtered separation negative. • These three halftone positives may now be utilized following standard photosilkscreen technique for three-color halftone work. If you intend to use a fourth printer (a separation negative for black), then reduce the above angles to approximately 15° and add the black at a 45° angle counterclockwise. Print the separations in this order: yellow, magenta, cyan, and black (optional).

Kodak Autoscreen Ortho Film 2563 can also be used for producing a halftone dot pattern. This film is especially useful where extra-large halftone separations are desired. • First, take the original three separation negatives and make three positive 20x25cm (8x10in.) prints. These will look gray and somewhat flat but will exhibit substantial detail. • Next, center each print on a separate but same-sized piece of white mat board in such a way that all images read in the same direction and are parallel to the base of the mat board. Permanently tape down the first separation print (blue-filtered). • Pivot the second separation print (red-filtered) from its center approximately 30° clockwise and tape it down. • Pivot the third separation print (green-filtered) from its center approximately 30° counterclockwise and tape it down. • Using a 4x5 camera loaded with Kodak Autoscreen Ortho Film (133-line), photograph each of the mat boards in turn with the lighting and copying technique recommended earlier for artwork and color prints. These negatives, after processing, are then enlarged to the desired positive size on high-contrast litho film.

During enlargement, the dots will increase in size as well as in the distance of each dot from another, so that the dot pattern becomes coarser. A dot pattern in the range of 70-line to 30-line is suitable for photosilkscreening.

If a still larger dot pattern is desired, increase the distance between the separation print and the camera, so that the image takes up less space on the Autoscreen film. If a tighter dot pattern is desired, decrease the distance between the camera and the separation print, or refrain from enlarging the separation negative so greatly.

If one does not have access to 4x5 camera equipment and a reasonably small three-color reproduction is desired, use an enlarger to project 35mm separation negatives onto 50mm squares of Kodak Autoscreen Film 2563, being careful to angle the film properly on the baseboard before each exposure. These screened and angled positives are then contact-printed onto high-contrast litho film to make separation negatives, which can then be enlarged to make three screened positive separations suitable for photosilkscreening.

Printing pigments should be mixed and balanced in the complementary colors of the filter used. One will need a cyan, a magenta, and a yellow paint. There are commercially prepared halftone-process colors available in black, blue, magenta, and yellow (e.g., Naz-Dar Halftone Colors). They can be cut with 20% to 25% halftone clear base to facilitate color balancing. A few drops of all-purpose oil can also be added to curb a rapid drying cycle. For general work, use a monofilament polyester mesh and misalign the dots slightly when printing, in order to insure good coverage. (See Chapter 9, photosilkscreen technique.)

Photosilkscreen Technique

Photosilkscreening is a useful tool for transferring photo images to a multitude of surfaces, including paper, metal, textiles, glass, and ceramics. Silkscreening is more time-consuming than direct application of a photosensitive resist or pigmented emulsion; however, there are real cost benefits, in terms of dollars and of time saved, when longer runs are undertaken in production of a single image. The silkscreened image also has a unique character of its own, creating a slight relief effect on the surface of an object. One will find that there is a wide variety of inks, paints, resists, and home-brewed concoctions that can be applied to meet almost every need. Technical back-up information and materials can readily be obtained from local silkscreen dealers. Here, we shall limit ourselves to exploring the methods by which light-sensitive materials are used to convert a piece of fabric into a photographic stencil for printing.

FABRICS FOR SCREENS

Silk, the fabric of multifilament construction that gave the process its name, is slowly being replaced in "silk" printing screens by less expensive and longer-lasting manmade fibers. Synthetic multifilaments, such as Dacron, offer a stronger, more uniform weave. The monofilament materials, such as nylon, polyesters, or wire cloth, offer even greater uniformity in the weave and so allow for a freer passage of the ink. Generally, synthetic fabrics are the best for liquid screen emulsions. These emulsions are used to coat the screen in order that a photographic stencil can be created directly on the mesh. If one would prefer to use indirect stencil film, then the synthetic mesh must first be treated with a special bonding agent to improve adhesion.

For optimum printing results, it is always advisable to choose the fabric to suit the task; however, a monofilament weave can be recommended as an excellent multipurpose material for most applications in this book. Monofilament polyester is preferred over nylon, which has a tendency to stretch, or stainless steel, which is expensive and tends to deform. A 200-mesh monofilament poylester will hold either a fine-line or a halftone image well. Alternative fabrics suitable for high-contrast photographic images are a 12xx to 14xx silk or multifilament polyester. Mesh count might be viewed best in terms of image sharpness: Generally, the tighter the mesh, the less the amount of ink that is laid down on the object during printing and the finer is the resultant detail in the image. Coarse fabrics tend to lay down too much ink and are preferable for bold, graphic photo images and some textile printing applications. One can employ a mesh count as fine as 16xx or 245-mesh for photographic work, but a tighter weave is not usually recommended. 200-mesh monofilament polyester is comparable to a 12xx multifilament fabric in regard to ink deposit, yet it will yield a sharper image without the problems associated with tighter weaves.

PREPARATION OF FRAME AND SCREEN

When building a screen, it is advisable to make it reasonably large, yet easy to manage. Each time a new screen is made there is a high wastage factor, and good fabric is expensive. A useful screen size is one that can accommodate two 41x51cm (16x20in.) photostencils. This size is suitable for doing a three-color 20x25cm (8x10in.) print on one screen. When measuring out the mesh required, calculate 5cm (2in.) extra along each side of the largest stencil that will be made on the screen, plus 5cm extra per side for tacking and ease in stretching the fabric, a total of 10cm (4in.) extra on each side. For example, to build a screen that would accommodate a 41x51cm (16x20in.) stencil you would need a piece of fabric 46x56cm (18x22in.), plus another 5cm (2in.) all around for stapling and grasping, or a 51x61cm (20x24in.) piece of material. In this instance the inner dimensions of the screen frame would be 46x56cm (18x22in.).

Framing material can be obtained from a silkscreen store. These frames have interlocking corners and are grooved to accept a stretch cord. One can also make frames from 2x2in. (5x5cm) kiln-dried wood. Cut the 2x2's and join the pieces at the ends with lap-joint construction. Glue and nail the corners for added strength. Work on a level surface to prevent warpage, and make certain all corners are exactly square.

To fit the mesh to the commercially made frame, first join the frame lumber together, then use a knife to round off any sharp edges at the corners in the groove. Lay the mesh over the frame. doubling it at each corner. Insert a tack through the doubled material at one corner. Pull the mesh as taut as you can without tearing it and insert a tack at a second corner. When tacks at all four corners are keeping the fabric under tension, check that the holes in the mesh are in alignment with each other. Now bind the mesh to the frame with cord: Lay a continuous strand over the mesh along the groove around the frame and with a hammer and a blunt object (such as a door hinge) drive it halfway into the groove. Remove the corner tacks and finish inserting the cord. If the groove is uneven in width, the cord may lift out. This problem can be remedied by wrapping a length of cord several times in some leftover mesh and wedging it

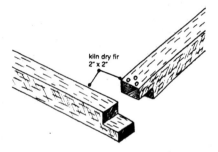

Lap-joint construction; glue and nail together on flat surface.

kiln dry fir
2" x 2"

into the groove on top of the first piece. The frame is now complete and can be sealed with tape.

To stretch the mesh on the homemade frame, begin by stapling (use a heavy-duty staple gun) from the center of each side to create even tension. Staples are inserted in either the top or the outer sides of the frame, at a 45° angle, and into staple tape (optional) in order to avoid runs in the mesh. Fasten one or two staples at the middle of one side, then go to the opposite side, stretch the fabric (being sure it is very straight) and place one or two staples in tape in the middle of that side. Then go to first one end and then the other, tensing, straightening, and stapling the fabric as described above. Returning to the first side, place one or two staples on each side of the center staples, no more than 13mm (½in.) apart to insure even tension. Next do the same on the opposite side and then on each end. Slowly the work will progress outward to the corners, following this similar pattern around the frame, tautening the fabric as you go. All the corners should be reached at approximately the same time. Silk and polyester will stretch only so far before they tear. Nylon, on the other hand, seems to want to stretch forever. This is why nylon is recommended for printing on curved and irregular surfaces. To make nylon extra taut, it should be presoaked in water and stretched wet, since it has a tendency to loosen up while it is moist.

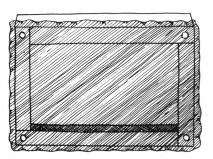 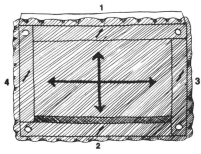 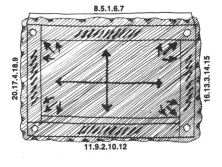

After the screen has been stretched and stapled, run a finger lightly over the mesh around the edge of the frame on the outside. This is the side that will be flush and in contact with the printing surface. The lipped side (the inside) of the stretched frame is the side where the ink will be applied and squeegeed through the mesh. The mesh should feel reasonably smooth, with no rippling evident. If there are ripples, the screen has not been stretched properly. In this case, try adding a few more well-placed staples. A coating of gloss acrylic medium is now applied to all sides of the framework, plus 13mm (½in.) inward on the screen's working surface. The application of medium tightens up the screen considerably. Next, apply overlapping strips of 5cm (2in.) packaging tape on both the inside and the outside of the framework. The frame should now be completely covered with tape. The tape should extend no more than 13mm (½in.) into the working area of the screen. The tape is then sealed with two or three coats of acrylic medium. The tape and the medium in combination waterproof the frame, preventing warpage. They also help keep the ink from becoming lodged between the

How to stretch a silkscreen: (left to right).
1. Temporarily tack down mesh at all four corners under minimum tension. This holds the material in place on the frame. 2. Place staples 1, 2, 3, 4 at a 45° angle into the frame. Black arrows represent direction of tension while stretching the fabric. 3. Proceed stapling and stretching from the middle of each side to the corners. Try and arrive at all corners approximately the same time.

mesh and the framework, making cleanup between color changes easier. One can also seal the edges of the screen by coating them with a sensitized waterproof screen emulsion, permanently sealing the edges via exposure to UV light. This is a good technique for use with photostencil film only.

In order to use the screen most effectively, a printing board is required. Use a piece of 19mm (¾in.) exterior plywood, 5cm to 8cm (2-3in.) per side larger than the screen itself. Purchase a pair of inexpensive pin hinges found at hardware stores. Attach the hinges to the frame and to the plywood (or hinge bar) so that the screen will center on the printing board, flat side down. Secure hinges to a long side of the screen and the printing board. Place the hinges close enough together so that they can accommodate screens of varying sizes; however, be careful not to space them so close that "screen-wobble" becomes a problem during printing. With a pin-hinge system each screen frame has its own set of two half hinges and the printing board has a master set of opposite interlocking halves to accommodate the different frames. A frame is joined to the printing board by inserting a pin between the aligned half hinges. Insert the pins in opposite directions to minimize play in the hinges. There are also other types of hinging systems available. One such alternative type consists of commercially available screen hinges. The C-clamp type of hinge attaches to the printing board, will accept frames of different thicknesses, and does not require separate hinge sets for each frame. Finally, a temporary hinge of masking tape can be used in an emergency to hold smaller screens to any flat working surface.

During print runs, it is useful to elevate the screen off the board to accommodate paper changes and registration of multicolor image segments. A simple solution to this problem is to fasten a 15cm (6in) piece of dowel, acting as a leg device, approximately halfway along one side of the screen frame. Use a single nail or screw so that it will pivot.

Before a screen is used, it must be degreased so that emulsion will adhere to it. Wet down the fabric and apply a solution of trisodium phosphate (TSP) sponging it liberally onto the surface of the screen. This degreaser has no abrasives in it and will aid in removal of any oil or sizing in the mesh. To mix a solution, stir about 30ml (2 tablespoons) of TSP into 900ml of warm water. Household cleansers (such as Ajax) may also be used, but they are more abrasive. In addition there are several commercially prepared screen-fabric degreasers available. Once the screen has been degreased, rinse it well in water to remove all solutions and any particles. Wipe it with a soft rag soaked in vinegar or a weak acetic acid solution (4%). In order to use the screen, all that remains is the application of the photographic image, in stencil form.

METHODS OF APPLYING THE IMAGE TO THE SCREEN

There are two established methods for applying a photographic image to a silkscreen: The indirect stencil film method and the direct screen emulsion method. The indirect method requires a stencil film (a layer of light-sensitive emulsion coated on a temporary support) for image formation. The image is photographically created via contact printing, special development, and washing out. After processing, the image in stencil form is transferred to the screen and separated from its plastic support. Stencil films are useful where oil-based or lacquer-based inks are employed as the stencil material is resistant to solvents, yet easily removed with water. Different thicknesses of stencil film are also available, the thinner films affording better detail as required for fine line or halftone work. These thinner films, however, are not as wear resistant as the thicker films and will not hold up for long print runs without some image deterioration.

The direct method of screen image formation uses a photosensitive liquid to coat the mesh. The image in stencil form is created *directly* on the emulsion-impregnated screen via contact printing technique and wash out procedure. Direct emulsions are available for use with either water-based or oil-based inks. These emulsions are also well suited to halftone work. The only disadvantage with direct emulsions is that stencil removal is sometimes difficult. As a general comparison, either means of stencil fabrication will yield good results when the materials are matched to the mesh and the screen application.

Both of the above methods require a positive transparency to produce a positive print (review Chapter 8). If one plans to screen colorful posterized images, several positive transparencies of different colors and/or densities (shadows, midtones, highlights, etc.) will be necessary, and a photostencil must be made from each. By utilizing butcher paper and masking tape, however, one can mask off portions of an image, thereby obtaining several image variations and corresponding color variations from just one photostencil. One might also try printing a second color through the same stencil, but with the image slightly out of register.

THE INDIRECT STENCIL FILM METHOD

The indirect method requires the application of a photostencil film to the screen. There are many stencil films on the market, and they all work basically the same way. To begin experimentation with photostencil film, try McGraw-Colorgraph Type 4570 or one of the Ulano films, Blue Poly or Hi-Fi Green. These products are available from most silkscreen supply dealers. Type 4570 is unique in that no special developer is required for processing, other than water. A dilute hydrogen peroxide solution may be used, however, to obtain finer detail, especially on outdated film. (Use one part 3% hydrogen peroxide to eight parts water.) With Ulano photostencil films, Ulano "A" and "B" Powder Developer is normally recommended. It is packaged in premeasured pouches and produces exceptionally good results. Where image detail is not so critical, try using an ammonium dichromate or potassium dichromate bath as a substitute developer. (Mix 14 grams per liter of water.)

The processing steps for the indirect method are as follows:
1. Exposure. All stencil film should be handled under subdued light. A low-watt yellow light placed about three meters (10 feet) away from the work area should give adequate visability, yet still prevent fogging of the film. Cut a piece of stencil film about 2.5cm

Exposure system for Indirect photostencil film.

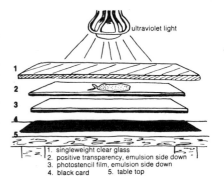

1. singleweight clear glass
2. positive transparency, emulsion side down
3. photostencil film, emulsion side down
4. black card   5. table top

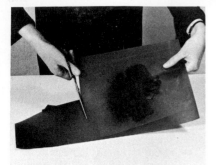

Indirect photostencil method for silkscreen application.
This page, left to right from top:
1. *In subdued light cut a piece of stencil film 1 inch larger than the transparency.* ***2.*** *Place stencil, emulsion side down, on a piece of black card. Lay positive transparency on top, emulsion side down—sandwich together with sheet of glass.* ***3.*** *Expose to an ultraviolet light source.* ***4.*** *Develop in Ulano A & B Developer for approximately 90 seconds.*
5. *Wash the image, emulsion side up, in hot water until all the unexposed areas are removed. Next chill the film in cold water to harden up the remaining gelatin.*

(1in.) larger all around than the positive transparency you plan to reproduce, and place it emulsion-side down on a piece of dark, light-absorbent material. The exposure is made through the acetate backing of the stencil film. It is important to remember that the stencil's emulsion side will be affixed to the flat side of the screen, causing the image to reverse from how it appears during exposure. Therefore, place the positive transparency on top of the stencil with the left side of the image at the right. For example, any writing must appear backwards during exposure, if it is to reverse properly. Place a piece of thin, clean glass on top. The glass should be large enough to cover the film in its entirety and heavy enough to maintain very even contact between the transparency and the stencil film's acetate backing. In this manner optimum image sharpness is possible. Other alternatives for ensuring good contact during exposure include the use of a contact printing frame or vacuum frame.

Exposure is made with any ultraviolet light source at a distance of at least the diagonal of the transparency. A test strip exposed in one-minute increments is a useful tool for determining the best times. A typical exposure using a black light tube is about seven minutes at 46cm (18in.) distance. It is the action of light passing through the stencil that hardens the emulsion. Not enough light (underexposure), and important image detail will wash out; too much light (overexposure) and even the highlights will block up, preventing ink from flowing through the stencil. One will find that there is generally more latitude toward overexposure.

2. Development. After the stencil film has been exposed, immerse it in the appropriate developer solution. Agitate the film in its developer for about 90 seconds, emulsion side up. Remember that Ulano

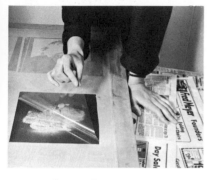

Indirect photostencil method for silkscreen application. (continued) **6.** *Lay the stencil, emulsion side up, on paper towels and newspaper. Position screen over image.* **7.** *Lay screen down gently and blot image with paper towels to remove excess moisture and ensure optimum contact. The gelatin actually impregnates the mesh holes.* **8.** *Turn screen over and apply block-out medium where required. Brush right over the backing sheet.* **9.** *Once the film is completely dry, the plastic backing can be easily peeled off. Start at one corner and proceed slowly.* **10.** *A thin paper stencil mask can be applied to the printing surface with tape. The initial inking will help to adhere the paper about the image. This is one alternative to block-out medium for print runs of short length.*

"A" and "B" developer forms a strong gas upon being mixed together. It must not be bottled and stored at the end of a work session. If one is processing several sheets of stencil material over a few hours, then cover the developer tray with an opaque lid between usage. This will protect the solution from direct light, which can spoil it prematurely.

3. Wash Out. Next immerse the stencil, emulsion side up, in a bath of hot water (43°C; 110°F) for about five minutes. The hot water removes the unexposed areas of the stencil. Keep rinsing it under the water until all the bubbles are gone. Be careful not to touch the stencil at this time, as its surface is very easily scratched and damaged. Once the image is completely developed, rinse the stencil in cold water. This sets the image by hardening the gelatin emulsion.

4. Attaching the Stencil to the Screen. If a synthetic screen material is used, the mesh must be conditioned first with a commercial bonding agent. This permits the stencil to adhere better to the slick synthetic fibers. A bonding agent is not necessary for applying a stencil to silk. Prior to application of the film, premoisten the screen fabric to make it more receptive to the stencil.

The stencil, while still wet, should be placed emulsion side up on a flat surface covered with newspaper. Position the screen, flush side of the frame down, directly over where you want the stencil and gently lay the screen down on top of it. Without moving the frame, cautiously blot the screen with paper towels or newsprint to remove excess water from the inside (the inking side) of the screen. This will insure a good contact between the stencil and the screen. After the stencil has set for 10 minutes, stand the screen upright to dry overnight. A fan can help speed the process considerably, but heat and uneven drying are to be avoided. It is best to leave the acetate backing on the stencil until the screen is ready to be used in order to protect the image from possible damage. The backing will peel off easily, offering no resistance, when the screen/stencil sandwich is thoroughly dry.

5. Preparations Before Printing. This is the time to touch up any pinholes in the stencil and to block out the surrounding open areas of the screen to keep ink from flowing through the mesh where it is not wanted. Eliminate larger, open areas of the screen with wide strips of masking tape and butcher paper or a hand-cut newsprint stencil. An excellent blockout medium for smaller areas is Water Sol. Apply this quick-drying liquid to both the stencil side and the inking side of the screen. This will help insure that ink does not settle on the blockout solution and dry in the mesh, making it especially difficult to clean the screen. Use a brush or a piece of stiff cardboard to apply a light, even coat. Be careful not to get the filler on the stencil itself, or it will plug up the image. If the acetate backing is left on the stencil during application, this problem can be minimized. Once screen masking is completed and the medium is dry, one can peel the acetate backing off the stencil and begin to print.

6. Printing. Review the last section of this chapter for general hints and suggestions.

7. Cleanup of an Indirect Stencil. Lay the screen on top of several sheets of newspaper and liberally apply paint thinner. Use lacquer thinner (wear gloves) for stubborn paint removal. Wipe up the excess with paper towels. Reapply thinner until all paint is removed. If one intends to reuse a stencil, it is only necessary to clean the paint from the image area before another color may be applied.

To clean the screen for reapplication of a new stencil, hose it down with water. Cold water will remove the blockout medium and hot water will remove the photostencil film. Again, flood the screen with solvent to remove the last paint. Lingering bits of paint are quickly removed from the mesh by rubbing both sides with rags, such as discarded T-shirts, soaked in solvent. A faint image may remain on the screen, but none of the holes should be blocked up. Commercial stencil removers and screen cleaners are available and might be useful, especially in the final cleaning stage. Once the screen has been cleaned it can be stored or reused immediately.

THE DIRECT SCREEN EMULSION METHOD

There are two kinds of sensitizer available for activating a direct emulsion; one is a diazo type and the other consists of an ammonium dichromate solution. Although packed differently, both sensitizers are used with a similar emulsion base. This base appears to be a polyvinyl-alcohol/polyvinyl-acetate blend, sometimes with a dye concentrate added for increased visibility. PVA/PVA colloids are water-soluble resins that, when mixed with either diazo or dichromated sensitizer, will harden upon exposure to UV light.

Diazo type emulsions exhibit excellent keeping properties, in comparison with dichromated direct emulsions, once they are mixed. Diazo emulsions permit screens to be coated and stored for a much as three months prior to use. Dichromated emulsions, on the other hand, should be used soon after the emulsion and sensitizer are mixed and/or the screen prepared. Another important difference is that exposures with dichromated emulsions tend to be twice as fast as those experienced with diazo-type emulsions.

There are several good diazo-sensitized brands available, such as Naz-Dar Encosol-3, a direct emulsion to be used with water-based inks, or Encosol-1 and -2, suitable for use with solvent-based inks. If one wishes to try a dichromated emulsion, purchase a quart of #71 Screen Star Photo Emulsion (regular viscosity) or a comparable brand available from silkscreen suppliers.

The major steps for using either a diazo-type or a dichromated screen emulsion are outlined below:

1. Mixing the Emulsion. Diazo emulsions are packaged with sensitizer, emulsion base, and dye concentrate. First completely dissolve the sensitizer by filling up the balance of its container with water. Pour the sensitizer into the emulsion base and blend well with a plastic rod. In order to increase image contrast, making it more visible after wash-out, the dye concentrate can be added. It is best to add only a quarter of the dye provided, as the exposure times may be substantially increased with larger amounts. The diazo emulsion is now ready for screen application.

Dichromated emulsions are usually purchased as a tinted emulsion base. The ammonium dichromate salts are purchased separately. To make a saturated sensitizer, mix 28 grams of ammonium dichromate with 240ml of water. Just prior to use, slowly stir in one part of sensitizer to five parts of emulsion base. The dichromated emulsion may now be applied to the screen.

As a substitute for a commercial emulsion, a sensitized gum print solution can be used for short runs with oil-based inks.

2. Applying the Emulsion. To apply the emulsion, an applicator of some sort will be necessary. An emulsion applicator can be made from a piece of plastic or a stiff piece of cardboard. It is not essential that the applicator be able to cover the image area in one stroke. A thin, even coating of the emulsion is applied on the outside (flush side) of the screen, followed by a thin application on the inside (inking side). Lightly scrape off any uneven excess that may remain on either side of the screen. The objective is to fill all the holes in the mesh with emulsion, while maintaining as smooth a coat as possible, particularly on the outside of the screen.

Dry the screen, preferably in a horizontal position, in a dust-free place away from light. With diazo-type emulsions, a hair dryer can be used to expedite drying without danger of heat fog. Should pinholing be evident, apply an additional thin coat of emulsion to the outside (flush side) of the screen. Once the emulsion starts to dry, it becomes light-sensitive and should be handled under subdued light until water-development.

3. Exposure. Prepare the screen for exposure by placing it with the inking side (the rimmed inner side) of the mesh facing downward, resting on a 6cm (2½in.) foam pad. This pad will act as a support under the mesh during exposure. Between the foam pad and the screen fabric, place a black, nonreflectant material. The positive transparency is now positioned on top of the sensitized mesh. If a halftone transparency is being utilized, turn it to an approximate 22½° angle out of parallel with the stretched mesh in order to avoid a moiré effect. (The moiré pattern is made visible by rotating a halftone transparency over the mesh surface. It looks like an explosion of stars.) This precaution insures a superior ink flow through the screen with less bleeding together of the dots. Another alternative is to prestretch the mesh at a 22½° angle to the sides of the screen. In this way, screen space is utilized more efficiently, physically permitting more halftone transparencies to be printed on a comparable mesh area.

As mentioned in the Halftone Preparation section of Chapter 8, a 40-line to 65-line screen is advised for making the halftones, although a slightly finer screen may be used. Too minute a dot pattern will neither hold well on the mesh nor print satisfactorily. The dot pattern from a coarse screen will be visible on the printed surface; however, this is not necessarily objectionable.

If the work is very critical, first make a photographic test strip on a sample screen or a small piece of white cardstock, in one-minute increments, to establish an exposure range. There is a healthy mar-

Direct Photoemulsion method for Photosilkscreen (left to right from top):
1. *Coat the mesh in subdued light with sensitized direct photoemulsion.* ***2.*** *Dry the coating in the dark. A fan can assist the drying operation.* ***3.*** *Exposure technique for direct photosilkscreen emulsion.* ***4.*** *Unexposed image areas are washed out with a mild spray of warm water.* ***5.*** *Use block out medium or sensitized direct emulsion to spot pinholes. Dry the screen and print.*

1. singleweight sheet of clear glass
2. positive transparency, emulsion side down
3. sensitized photosilkscreen, inking side down
4. black, light absorbent card stock
5. 2½" foam pad   6. table top

gin for overexposure. A sun lamp at approximately 64cm (25 in.) for 15 minutes gives very good results, although any UV light source will work. Sometimes it may be necessary to raise the light source from a known exposure height in order to insure a broad enough illumination. This alteration will call for adjustments in known exposure times. As a general rule, multiply a known exposure time by four whenever the distance from the light source to the transparency/screen sandwich is doubled. This should give one an approximate time for the new exposure. After an exposure time has been established, double check to make certain that all image components will "read" correctly when the screen is inverted for printing. Finally, lay a piece of glass, larger than the positive transparency employed, directly over the film/sensitized-mesh sandwich, to ensure tight contact during exposure.

4. Development. After the screen has been exposed, unexposed areas are washed out in a water development. Place the screen under a soft spray of warm water, increasing the pressure and water temperature (43°C; 110°F), after one minute. Wet both sides of the screen. Where lack of space prohibits spraying, use several bucketsful of warm-to-hot water to loosen the emulsion. When the image appears sharp and further "development" has no effect, blot the mesh on both sides with paper towels and allow the screen to dry. A fan may be used to expedite matters. Next, mask off those areas of the screen that are not to be printed, using masking tape and butcher paper. If any touchup work needs to be done, use some of the emulsion and let it set in UV light. When masking and touch-up is complete, the screen is ready to be printed.

Occasionally an emulsion is given a hardening treatment to increase resistance to humidity, water-based inks, and strong solvents. Emulsion hardeners are applied by brush directly after water "development." Once treatment is completed, the screen is set aside to dry.

5. Printing. Review general suggestions on printing in the last section of this chapter.

6. Removal of Direct Emulsions. After the ink has been removed, the screen may be stored indefinitely or reclaimed. If a direct emulsion is used on silk material, the fabric cannot be salvaged. To reclaim a screen, soak the mesh in a full-strength solution of household bleach for approximately 15 minutes. Take several 5x10cm pieces of wood (2x4's) and create a rectangle large enough to accommodate the screen. Lay a piece of plastic sheeting over the wooden frame, forming a basin to hold the bleach and the printing frame. This method permits one to save and reuse the bleach. You can also lay several paper towels or newspapers under the screen, printing (flat) side down, and pour the bleach over the mesh. Next, while scrubbing with a soft brush (e.g., a toothbrush) dipped in bleach, use a hot water spray to "blow out" the loosened emulsion. Sometimes the emulsion really holds tight. For those stubborn spots, use the following procedure: Identify them by circling each spot with a pencil directly on the mesh. Place paper towels under each spot and pour on some bleach. Next pour on a similar amount of 3% hydrogen peroxide. (Note: wear rubber gloves and beware of toxic fumes.) Let this mix set for 20 minutes, then scrub and blast with water again. As a last resort, use a high-pressure spray gun like those found at most U-Wash car facilities. (Emulsions tend to set harder with age; therefore, remove the emulsion as soon as possible after use to avoid problems like this. Use one week as a guideline.)

PROJECTION-SPEED SCREEN EMULSION
The Rockland Colloid Corporation of Piermont, N.Y., manufactures and markets an enlargement-speed screen emulsion called SC-12. It

is applied in the same manner as any other direct emulsion; however, here one can use a camera-size negative and a photographic enlarger as the light source to expose the emulsion. This product can be useful for making extra-large screens where an oversized transparency might not be available.

Liquify a portion of the emulsion and apply it under red safelight while it is still warm (35°C; 95°F). Once the emulsion has dried, the screen can be exposed. After exposure, sponge SC-12 developer onto both sides of the screen. (It is advisable to wear rubber gloves.) The unexposed image areas are next removed with a hot water rinse (46°C; 115°F). This completes development, and the image is set by rinsing it with a mild bath of acetic acid. Once the stencil has dried, it may be printed using latex paint or any oil-based ink.

After use, synthetic screen mesh can be reclaimed with any household bleach. Silk mesh cannot be salvaged.

SC-12 can also be employed as a photoresist on certain metals. This will necessitate some experimentation according to each individual's specific requirements.

HOW TO MAKE A VACUUM TABLE

A vacuum table can be an invaluable asset to any photosilkscreen operation. It will help to keep the paper or material flat and in register during print runs. Simply described, it is a special printing board to which the silkscreen frame is attached, permitting the fast and accurate registration of print materials. It works by creating suction through the tiny holes over the board's surface, holding the material in place. Screen height can also be varied so that materials of varying thickness can be accommodated.

A vacuum table can be made easily as follows:

1. Secure a piece of 2.5cm (1 in.) thick pressboard, the length and width dependent upon the size of the printing area desired. Plywood is not recommended.

2. Gouge a channel all the way down the middle of the board (North-and-South), approximately 3.8cm wide by 1.3cm deep (1½x½ in.).

3. Simliarly rout out channels running perpendicular to the center channel (East-and-West), 6mm wide by 6mm deep (¼x¼ in.). Make each channel 2.5cm (1 in.) apart, measuring from center to center. Start each about 5cm (2 in.) in from each side and run through the main channel. Begin routing these smaller channels about 12.5cm (5 in.) down from the top of the board (North) and stop approximately 5cm (2 in.) from the bottom (South).

4. Place a piece of tracing paper the same size as the pressboard over the channels and rub a pattern of their location with the side of a pencil.

5. Permanently secure a piece of laminated plastic such as Formica over the incised side of the board, using contact cement. The plastic should be the same size as the pressboard.

6. Use rubber cement to glue the paper pattern over the Formica On every row shown on the pattern, drill holes 25mm (1 in.) apart, so that there will be a 3mm (1/16 in.) round hole every inch (25mm) over the entire printing surface.

7. Insert a flat vacuum nozzle ("crevice" attachment) into the North end of the center channel. Seal it in place permanently with an epoxy glue. When the vacuum table is to be used, one need only attach the hose of a household vacuum cleaner to the insert. The opposite end of the table is fitted with a sliding door device (see illustration) to permit control of the suction on the table.

8. Drill two holes through the table at the North end. Cut a piece of 5x5cm (2x2 in.) kiln-dried wood, the width of the table (East-West measurement) and drill two holes in it to match up with those in the table. This piece of wood is called the adjustable hinge bar. The bar is attached to the top of the table with bolts extending up through the table from underneath. Wing nuts are used to tighten the bar down on the printing table. The bar can be raised or lowered to accommodate any size object or print by adjusting the wing nuts and inserting or removing shims (wooden wedges) under the bar. The shims should all be made the same size and thickness.

9. Fasten two master hinges, each consisting of one-half of a removable-pin hinge, permanently to the top of the bar and flush with the edge, to attach printing screens. Be sure to space them to match the pair of removable-pin hinge halves on each of your screens as the frame rests in its printing position. The hinges are placed so that the screen will lift up during print runs. (Review Preparation of Frame.)

Constructing a Vacuum Table: Below from left to right.
1. The routed piece of pressboard, ready to have a tracing made for the formica top. 2. Adjustable hinge bar raises and lowers off the table top with the aid of shims. 3. Permanently glue (A-2) into (A-1). (A-3) is removable. 4. Sliding door of metal or plastic helps to control the suction on the vacuum table.

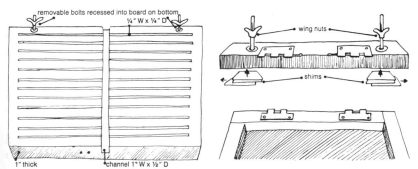

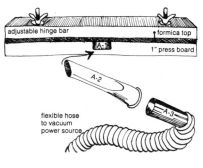

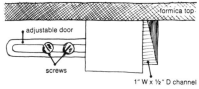

GENERAL PHOTOSILKSCREEN PRINTING TECHNIQUES

1. Positioning Images. Regardless of the type of emulsion system employed, there are some basic printing techniques that should prove helpful when screening the image. Some type of positioning, or registration, system will be required in order to insure accurate replication of work from print to print. One system requires the use of the original positive transparency used to make the photostencil. It is centered and taped down on a sample piece of the paper to be screened. Slide this composite under the screen, moving it about until it aligns perfectly with the stencil image. A few pieces of thin cardboard may now be taped down on the printing board at the corners of the sample sheet, creating a "lip" that will serve as a paper positioning guide. When the first sheet is removed, another identical sized sheet can be placed in position quickly and accurately.

If a multicolor print is being made, use the method outlined above to realign the image, so that each color will be in register with every other. This time, lay the second transparency down on a sheet

Dennis Bookstaber, Untitled. A brownprint on hand sensitized paper with photosilkscreened graphics.

of paper on which the first image was screened, taping it in a position corresponding to where the second color segment is desired. In multicolor or in three-color halftone separation work, it is advisable to misalign the images and/or the dots slightly to make sure that all areas of the white paper will receive ink.

An alternate system for multicolor work uses a clear acetate sheet as the registration guide. The sheet, larger than the image area, is taped along one edge of the print board. Pull an image on the acetate. Align a piece of printing paper underneath it. Fold the acetate back out of the way and pull the print. If one is using more than one stencil or screen, then clean the image off the acetate with the appropriate solvent, replace it under the second stencil, again tape down one side, and pull the second image component. Now, realign a print of the first image under the acetate bearing the second image until the images are in register. The second color is now ready to be pulled after the acetate has been folded out of the way.

Another variation, suitable for longer print runs, involves the use of three very short registration pins taped down on a thin strip of acetate the exact distance of a three-hole punch apart. All the paper to be screened is precut and prepunched. Place a piece of paper over the pins and center the paper under the first image component. Tape the acetate strip down and pull the first color. Clean up the screen, then place a dried print of the first color over the pins and insert the acetate strip again under the screen. Realign the second stencil over the first image component. Retape the acetate strip down and pull the second color. All future pulls will be in exact registration. Use this same method for each additional color.

2. Selective Printing. a. Water-based blockout solutions can be used in conjunction with waterproof direct emulsions and oil-based pigments to mask and print images selectively. Paint out with solution those areas of the image one does not wish printed. Dry the surface and pull the print. Next remove the paint with solvent and the blockout solution with water. Selectively reapply the medium, then register and rescreen the print. Several colors may be added to a print without preparing extra screens or photostencils. **b.** Tracing-paper stencils can also be employed for selective screening of images. Tracing paper is useful, as it is both transparent and hard enough to offer adequate ink resistance. Sharper lines will be obtained with thin tracing paper. To use this method, first pull a master print. Lay the tracing paper over the image and outline in pencil the area to be selectively printed. Cut out the shape and affix the stencil with tape to the underside of the screen. Ink and pull the desired color. Remove the stencil, and create new ones for other segments of the image. This method can also be used to create a three-color processlike print. For example, you want to screen a halftone image of a lady in a dress: The background, dress, and flesh tones can be pulled one at a time using a blank screen and this tracing-paper stencil method. The detail and shading for face and dress are added by pulling the halftone image over these base colors.

3. Keeping Mesh Off the Paper. During printing, the entire silkscreen frame should be elevated slightly at all corners. This is important. As the squeegee passes over the mesh with ink, it is desirable for the mesh to snap back off the paper or material being screened. The image produced is much sharper, with less blurring or smearing experienced, when ink flow and surface contact are both kept at minimum. Maintain approximately 6mm (¼ in.) between the mesh and the printing surface by using small cardboard or wooden shims. One will find off-contact printing particularly useful when screening heavier materials, such as glass and metal.

4. Applying Ink. For printing the image, a squeegee will be needed to force the ink evenly through the mesh. There are several different hardnesses and blade shapes used for screening; ask a silkscreen supplier for a squeegee to fit the particular application. For general use, a plastic squeegee blade of medium shore hardness (60 to 65 durometer) is recommended. Softer blades (45 durometer) are better for printing textiles, where a heavier ink deposit is wanted, while harder blades (75 durometer) give thinner ink deposits and sharper images. The squeegee should be at least 25mm (1 in.) longer than the width of the image area to be pulled.

Plastic squeeze bottles make suitable ink applicators for controlling the amount of ink placed on the screen. Lay down a small "worm" of ink along the top of the stencil. Pull the squeegee at a 45° angle in one smooth, continuous pass for optimum results.

One must be careful that the ink does not dry in the screen while one is changing the material to be printed. It may prove helpful to flood the screen with ink after each pass, to help prevent blocking of the mesh. Xylene is also helpful in keeping the screen mesh open when using oil-based inks, but CAUTION is needed and GOOD VENTILATION is absolutely necessary when you use it. There are also several products made commercially just for this purpose.

5. Drying Prints. Screened prints may be dried by laying them flat, stacking them in horizontal drying racks, or suspending them from a clothesline.

Lastly, a word of WARNING: Photosilkscreening can be a real high, especially if it isn't done in a well-ventilated room!

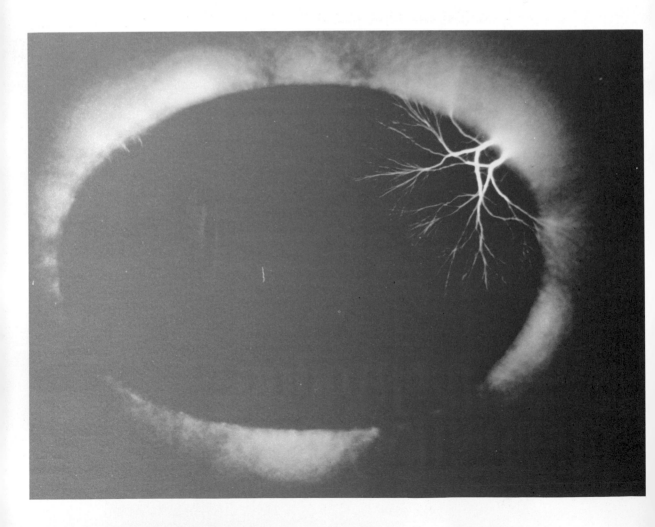

Other Imaging Systems

Newer forms of visual expression are continually being sought by the experimental photographer, and vast amounts of technology and of image-making tools are under consideration. There is already a variety of products employed daily in graphic communications that lend themselves to innovative application. In addition, new processes are being explored on the technical frontiers of photography. Several of these other image-making possibilities are presented here.

ELECTROSTATIC PHOTOGRAPHY: Images From Copier Machines

The office copying machine is being recognized increasingly for its artistic and photographic potentialities. It is an exciting visual tool that can be approached, not only as a printmaking device, but also as an instant camera capable of photographing three-dimensional objects.

Electrostatic photography is based on the premise that certain photosensitive conductive surfaces will accept and hold an electrical charge until that charge is dissipated by light or by other radiant energy. In respect to the office copier, the photoreceptor usually consists of a paper base coated with zinc oxide (direct method) or a selenium-coated intermediate drum (indirect method). The image or object, comprising dark areas and light areas, is reflected onto a charged surface, leaving a latent facsimile of the dark and the light areas. A pigmented resin, or toner, in either a liquid carrier or powder form is given an opposite electrical charge, causing the toner to be attracted to the charged areas in such a way as to produce a visible image. In the indirect method (e.g., xerography), the toner is electrostatically transferred, first to the drum and then to ordinary paper stock. Depending upon the method, the image is made permanent either by fusing the powder to the paper with heat or by evaporating the liquid carrier away and leaving an image in carbon black. With either method, a positive image results in a positive facsimile quickly and efficiently. This electrophotographic method is the basis for a number of variations to the copier process as it is known today.

There are many companies in the U.S.A. and elsewhere marketing reprographic equipment with a variety of capabilities. Some of the larger organizations, such as 3M, SCM, Xerox, IBM, etc., are well established and have branch or dealer facilities in nearly every large town, where one can become better aquainted with the artistic potential of copier systems. Some of the things copiers can do are listed below:

1. Copiers are instant printing systems, as well as instant cameras. By utilizing the total capability of the machine, visual ideas and concepts can be readily worked out without the time lag normally associated with photographic darkroom processing. Copiers are especially useful for creating instant reproductions of some three-dimensional objects, textural materials, transparency overlays, photographs (either matte or glossy surfaced), drawn additions, and "found" imagery from newspapers and magazines. When using found imagery, however, one should be aware of the new copyright restrictions, which apply to electrophotographic reproduction. Some direct systems permit the operator to resensitize the paper and re-expose it to additional imagery. This is also feasible with the indirect system, which uses ordinary paper.

2. Black-and-white copier reproduction from continuous-tone photographs usually results in contrasty prints. A textured screen can be used, however, to give the appearance of a longer tonal scale. For optimum black-and-white continuous-tone reproduction from a photograph, try one of the copiers in the Xerox 3100 family.

3. The density of the image (light and dark shades), can be altered, enabling tonal separations to be made quickly with minor dial adjustments.

4. Some copiers permit the operator to copy on colored paper as well as on a variety of other paper stock. Graphic images work well in these applications.

5. Collages can easily be cut out, assembled, pasted up, and re-photographed on the copier's imaging area (platen). Portions of an image can be effectively blocked out by using white paper masks and rubber cement or a white opaquing liquid, so that foreground and/or background of the image can be altered selectively.

6. Multiple copies of images can be made quickly just by setting the appropriate dial. Some copiers also make electrostatic offset masters to run on an offset press (e.g., A.B. Dick Electrostatic Offset Masters).

7. With a Xerox copier, positive transparencies on acetate can be made directly from the original for photoetching and silkscreen purposes. The toner deposit tends to be thinner than the black silver deposit on high-contrast film; therefore the untraviolet exposures needed in these applications will tend to be shorter than usual.

8. The copier reproduction of images on unsensitized paper permits hand-coloring (watercolors, opaque oil-based paints, transparent photo-oil colors, etc.) or the combining of images with

cyanotypes, gum prints, etc., by means of chemical transfer and/or reimaging of the paper's surface.

9. One can make copies of copies of copies of copies. With manipulation of the density control, third- and fourth-generation images acquire a line-drawing effect.

10. Reduction copiers are available, whereby one can create postage-stamp-sized images from normal-sized photographs. Reduction capability could be especially useful for instantly and selectively controlling image size in the production of photographic collages.

THE COLOR COPIER

Machine art is not limited to reproducing images in just black and white. Color copiers are available, opening up a whole new realm of visual possibilities. They offer an expedient means for producing full-color imagery as well as nearly unlimited color variations. Naturally, these images can be color-altered even further by the addition of handwork in paints, dyes, watercolors, etc. Currently there are only two color copiers readily accessible to the artist, the 3M Color-in-Color Copier and the Xerox 6500 Color Copier. There are, however, several others in different stages of development.

The future of the 3M Color-in-Color machine appears uncertain at this time for economic reasons, but it is worth discussing because of its unique color imaging process. The colors of any original or object situated on the glass platen are initially filter-separated onto intermediate film (coated with zinc oxide) and then treated ("developed") with carbon black. On the opposite side of each of the three color-separations are special dyes: the red-filtered separation is coated in cyan, the green-filtered separation in magenta, and the blue-filtered separation in yellow dye. Heat is applied to the carbon-image side, causing the dye to sublime, or vaporize, from the other side of the film to the surface of any paper previously loaded into the machine. The rich colors produced are "velvety" and soft-edged in quality. Images can also be reproduced on special matrix paper for transfer to alternative surfaces (cloth, acetate, etc.) in a dry-mounting press. Finally, it is recommended that 3M Color prints be sealed with a light coat of spray fixative.

The Xerox 6500 Color Copier also separates the colors of the original through an internal filtering system. Unlike the 3M copier, the Xerox color image is reproduced with toners of thermoplastic powder. (The colors are still the photographic primaries: cyan, magenta, and yellow.) Each color separation is transferred electrostatically from the reflected original to an intermediate drum to the selected paper stock. This machine can be adapted (with a Xerox 6500 Color Slide Adapter unit) to make color enlargements up to 21x36cm (8½x14 in) in size from 35mm slides, using a Kodak Carousel Model 600 or comparable projector. 75mm (3 in.) gel filters can be used in a special attachment to the slide-adapter, further widening the color imaging possibilities. There is also a Xerox Color Contrast Control Accessory that greatly improves the color rendition of continuous-tone images. It is a halftone screening device that helps reduce contrast.

Xerox color images can be produced on almost any 20# to 24# paper that will travel through the copier's roller mechanism and follow the necessary bends in the machine. Suitable stock includes overhead-transparency materials, plain bond, and colored papers with some variety in surface texture. Smoother surfaces are chosen when optimum image definition and sharpness are wanted.

Also practicable are special "release" papers, such as Zellerbach 300-J, which enable one to reproduce images on alternative surfaces (wood, fabric, etc.) via heat-transfer technique (review images on fabric, Chapter 4). In addition, Xerox color images can be transferred chemically, using a weak acetone solution, lighter fluid, or similar mixture (review section on combining images with gum prints). This enables one to transfer image components selectively. Lastly, Xerox color images can be lifted using Liquitex medium. (Review alternative transparency fabrication methods, Chapter 8).

PHOTOLITHOGRAPHY

Photolithography is a form of printing technology that uses either a piece of limestone or a metal plate to reproduce an image on paper. It is traditionally a planographic process (flat-surfaced printing) and relies on the principle that water and a greasy ink will not mix. A photographic image is processed on the chosen surface in such a way that the image areas are made receptive to ink, repelling water, while the nonimage areas conversely attract water and repel ink. This chemical action is the basis of photolithography using a flat-bed press (direct method), photo-offset lithography (indirect method, described in next section), and the collotype, or photogelatin, process. These systems of printing are distinguished from photogravure, or photointaglio, printing, wherein the ink settles in depressions forming the image and is lifted out of the grooves by pressure; from photographic images produced directly on light-sensitive surfaces; from photo-woodcuts, where the photographically applied image is later carved in relief, inked, and transferred by rubbing or pressing; and from photosilkscreening, where a photo-stencil is created on cloth and the ink is squeezed through it.

A lithographer's stone or an aluminum plate may be hand-prepared by photographic means for lithographic printing, using the following procedure:

1. Make a high-contrast negative transparency of the image, using normal darkroom techniques but remembering that the image will reverse during the flat-bed-press transfer. If there is writing or left-to-right appearance is important, this should be planned for in advance. If a continuous-tone-like image is desired, make a texture-screened or a halftone negative transparency. Separate transparencies and stones or plates will be required for each color utilized in a multicolored reproduction. If necessary, use an opaquing solution to do any touchup on the transparency.

2. Thick ball-grain aluminum plates (0.012in to 0.015in) should be employed when printing on a flat-bed litho press. Thinner plates tend to buckle. Grained, presensitized plates are also available. With these plates one can by-pass the counter-etch and sensitizing stages. With unsensitized aluminum plates, it is necessary to clean the plate chemically prior to application of the sensitizer: Prepare a suitable bath by diluting 180ml (6 oz) of 99% acetic acid in 3.79 liters (1 gal) of water. Dip the plate in the solution and rinse it in water. Repeat the dip and rinse steps, then let the plate dry. The plate is now ready to be sensitized.

3. There are several sensitizer/development kits on the market that contain lintless wipes, sensitizing ingredients, developers, and asphaltum gum etch. Try the Polychrome Corporation Kit, which contains Wipon Sensitizing Solution 909/910 and Polychrome Plate Developer #305. Mix this sensitizer, a negative-working diazo type, in the ratio of ½ capful of diazo grains to approximately 29ml of wetting agent. In subdued light or under yellow safelight, pour a small portion of the solution into the middle of the stone or plate. Spread it quickly over the surface with a lint-free wipe or a fine sponge until it appears uniform and there are no pinholes evident. Dry the stone or plate. Because of the proposity of the litho stone, two to three coats of sensitizer will be required. Use the same procedure, drying between coats. At all times avoid finger marks on the printing surface, as they can reproduce along with the image.

4. Once the stone or metal plate is dry, place the negative transparency, emulsion side down, on the sensitized surface (a piece of clear glass will help to insure good contact, if a vacuum frame is not available). Make the exposure under any ultraviolet source. (A 500-watt photoflood with reflector can be used at 46cm (18 in) from sensitized surface for 10 to 25 minutes, or a mercury vapor light, quartz lamp, etc.) Exposures will vary with the light source and with the thickness of the coating.

5. Pour a puddle of developer #305 onto the exposed surface. The developer is a desensitizing lacquer that makes the nonimage areas (corresponding to the black areas on the negative) ink-repellent and water-receptive. For approximately two minutes, spread the developer with a damp sponge, adding more developer if necessary, until the image is completely visible. Rinse the stone or plate of excess developer under running water and drain the surface for a moment.

6. Pour a solution of asphaltum gum etch onto the wet stone or plate and spread the solution with a damp sponge or lintless wipe. This protects the nonimage areas from oxidation and improves the water-repellent characteristics of the image area. (If the surface oxidizes, it will tend to loose its ink-repellent ability and print ink.) Buff the surface dry with cheesecloth.

7. When you are ready to proof the stone or plate, rinse or sponge off the surface with water to remove the asphaltum gum etch solution. The appropriate greasy ink can now be rolled over the surface and the image printed with conventional lithographic printing techniques. Finally, it is also feasible to use a thinner metal plate, prepared as outlined above, on a photo-offset press.

PHOTO-OFFSET

The use of photographic techniques on plates of thin metal or sensitized paper has made it possible for multicolored lithographic images to be made more expeditiously on rotary equipment. Offset,

Facing page: Peter Pfersick, "Zone Paint, The Photographers Friend." A photo-sculptural piece of photo offset lables and cans. 10 cans each 9" x 6" Above: Backside of "Zone Paint...."

or indirect, printing technique requires that a photographic plate be wrapped around a cylinder, which as it turns is both moistened and inked. The image from the ink-receptive areas is rolled off onto a rubber blanket on a second cylinder. It is then transferred again to its final paper support, which is backed by a third cylinder that insures good image transfer while assisting more paper, in roll or sheet form, to be drawn continuously through the machine. Other colors are added by using additional plates and inks for each color needed. Special lithographic inks are also available for image transfer to cloth.

Photo-offset lithography thus uses a rotary printing press rather than the traditional flat-bed system. It is inexpensive and less time-consuming than some of those methods of printmaking that require slow and methodical reinking by hand for each, or almost every, impression. Although photo-offset is slower and more expensive than a color copier system for short print runs, it does afford high-quality reproduction and is particularly economical for medium-to-long runs.

There are several ways to employ the photo-offset system in image-making. **a.** The original color image (scene, photo, color photocopy, gum print, etc.) can be color-separated by filtration. These three or four process-color separations can be employed to make photographic offset plates for full-color reproductions. **b.** Black-and-white images can be tonally separated into many density variations or combined with several other images. Colors can be selected, by separation or by image segment, to suit the subject matter. Prior to the actual print run, color-proofing materials can be used to help determine the final color composition (see next section). **c.** Bold, starkly graphic effects for offset printing can easily be achieved by using high-contrast film to record the subject matter. Where a continuous-tone-like reproduction is sought, the image should be screened prior to making the offset plate. **d.** Another possibility is to combine the offset image with other printmaking alternatives (gum prints, photosilkscreen, photointaglio images, etc.), selecting from various printing surfaces and compatible imaging systems.

The major limitation to photolithography is access to the necessary equipment. Many companies now have their own in-house printshops, and one might explore the possibility of using this equipment after hours on a fee basis. Alternatively, if access to a printing press is limited, positive black-and-white filter-separated or tonally separated prints can be made and given to a local printer for color reproduction to one's specifications.

COLOR PROOFING SYSTEMS

Color-proofing materials were devised in order to judge the quality of color separations before going ahead with the actual press run. They are contact imaging materials, highly useful for previsualizing true color in the final image. Adaptation of these materials to the fine arts offers exciting new dimensions in photography. Several companies manufacture these materials, and they are available through printing suppliers. There are two methods used widely in the graphic arts: one using colored overlays and the other, lamination. A new method that employs wipe-on pigments will also be considered.

1. Color Overlays

ENCO Naps and 3M Color Key both offer overlay color proofing using light-sensitive transparent sheets of pigmented film. The sheets come in various sizes, in process colors and in supplementary colors, and in either negative-acting or positive-acting form. A negative transparency will create a positive overlay with a negative-acting film, while a positive/positive or negative/negative image is produced with a positive-acting film. ENCO Naps overlay materials have a slightly thicker polyester film base, which is less apt to wrinkle when a lot of handling may be required.

Left: Color proofing film-processing equipment and materials: Sunlamp, Color proofing film, developer and transparencies. Below (left to right): How to process color proofing films: 1. Expose the transparency/film sandwich under glass to an ultraviolet light source. 2. Pour a puddle of developer over the exposed film. 3. Use a very soft sponge to gently remove unexposed image areas. (It is advisable to cover the bottom of a ribbed tray with a smooth sheet of plastic in order to avoid damaging the film during process steps.)

4. Rinse in water to remove all developer. 5. Hang to dry. The image may now be sandwiched in glass, used as an overlay, etc.

Color-proofing overlay films are all processed in a similar fashion: A high-contrast transparency is placed in contact (emulsion-to-emulsion) with a sheet of the light-sensitive overlay film under a subdued red or yellow safelight. It is exposed under glass to any UV light source (a sun lamp will do), and processed in its designated developer. The process is completed by washing the film and hanging it to dry.

By making several tonally separated transparencies from a given image, you can reconstruct the image in any color desired, printing each transparency on a different sheet of color-proofing film, then laying them one on top of the other. As these materials are transparent, they lend themselves to being mounted in a light box or sandwiched between glass and integrated into a greater design. A modified sandwich arrangement could consist of several pieces of film interspaced with clear sheets of acetate, or several overlays could be spaced 6mm (¼in) apart within an illuminated module. These spaced variations tend to alter the light pattern as the image is viewed from different angles giving the subject depth, and even movement.

Color-proofing films can also be employed to create new images that can in turn be photographed in register on a light box. In addition, four-color separations can be made and sandwiched together, producing full-color transparencies for decorative purposes.

2. Laminated Proofing

3M Transfer Key and Du Pont Chromalin are both color-proofing systems that employ lamination, whereby several layers of light-sensitive film are permanently superimposed into a single sheet capable of being mounted on a variety of surfaces (glass, wood, metal, plastic, etc.). The Du Pont system is unique, in that it employs toners of colored powder to develop the image after a sheet of clear, light-sensitive Chromalin film has been exposed to UV light. When the film is sandwiched with a positive transparency, ultraviolet light eliminates the tackiness in the exposed areas. Dry, colored toner is then rubbed over the image, where it adheres to the sticky areas.

Different colors are added by layering additional sheets of unexposed light-sensitive film, one on top of the other. Each layer is laminated, exposed to a separation transparency and dry-toned with the appropriate color pigment. This process appears to be a technically refined version of the dusting-on method (review section on nonsilver processes).

3M Transfer Key uses a pigmented sheet of light-sensitive film, exposed and developed the way 3M Color Key materials are. To add additional colors or image segments, you must laminate, expose, and develop each layer of precolored film separately, superimposing one atop the other.

3. Wipe-On Proofing Materials

The Kwik-Print system (Kwik-Proof), manufactured and marketed by the Direct Reproduction Corporation (DRC) is a relatively new form of color-proofing and image-making. It was developed as an offshoot of the whirler-applied Watercote color-proofing system, modified to meet the requirements and applications of the artist-photographer. Kwik-Print utilizes presensitized water-based, wipe-on pigments suspended in a colloid. These colors are not dyes, and are said by the manufacturer to have good archival qualities.

A wide selection of colors is available, including process-color equivalents (lemon yellow, medium blue, magenta, and black). All colors can be intermixed to generate new colors. Color intensity can be altered, a pastel-like effect created, and/or a suitable consistency for airbrush application obtained, by adding Kwik-Proof Clear. As ammonium dichromate has been added to all color solutions, it is recommended that gloves be worn during coating and development steps. Kwik-Print colors have a shelf life of nine months; therefore, pigment should be ordered in realistic quantities. Four-ounce and one-quart sizes are available.

Kwik-Print images are best reproduced on matte, vinyl-polyester sheets which are available from DRC in various contrasts and sizes. Tightly woven synthetic fabrics (acetate, polyester, etc.) or other nonporous matte surfaces, such as matte-etched glass or grained metal, can also be used with reasonable success. On natural fibrous materials, sizing with starch or with water-diluted white glue (e.g., Elmer's Glue-All), will minimize color absorption and process-related problems. On glossy surfaces, a "tooth" should first be created. Experimentation with several surfaces reveals that images can be made on metal, plastic, and papers (such as Strathmore's Bristol 100% rag paper) that have been matte surfaced with a retouching lacquer spray (e.g., McDonald Pro-Tecta Cote). Images exist on the thin lacquer film and are fragile when compared to those created on Kwik-Print materials, thus handle matte lacquered surfaces with extra care during processing. The final image can be protected with a glossy plastic spray, (such as Grumbacher's Tuf-film), which also tends to enrich the colors.

To make a Kwik-Print image, follow this procedure: Tape the back of the plastic sheet or other substratum down to a smooth, flat surface, such as the bottom of an unribbed plastic darkroom tray. Working under low lights, mix the pigment well by shaking the bottle, then pour a puddle into the middle of the sheet. Pour out only enough to cover the surface effortlessly. A puddle the size of a quarter is sufficient to cover a 20x25cm (8x10 in.) sheet.

Immediately after pouring the emulsion on the sheet, quickly, yet gently, spread it about the surface with a lintless wipe, (wrap the wipe around a smooth block of scrap wood, comfortable to the hand). Spread the color in the horizontal, then in the vertical direction, changing wipes as necessary. Continue to wipe the sheet until the surface is completely dry. While lintless wipes are preferred, a sponge or polyfoam brush can also be used to apply the pigment.

Once the pigment has dried, the image can be exposed. Exposures are relatively short and can be made with any UV source. For example, a sun lamp at 38cm (15 in.) requires exposures from one to four minutes, depending on the color used. Blue needs the shortest exposure and black requires the longest. The image is not visible until placed in water.

Flush polyester sheets under a faucet or hose (good water pressure is needed, however, tray development is recommended for the more fragile surfaces). The image will begin to appear within seconds. Development is helped by wiping the surface with a clean wipe. If the shadow areas are slightly overexposed or the whites veiled with a tinge of color, then pour a weak ammonia solution over the sheet, wipe, and water-rinse. (A capful of household ammonia to a liter of water, or 6ml of 28% ammonia to 4 liters of water, is usually sufficient.) Kwik-Print Brightener is a stronger reducing agent and is useful when an ammonia bath is unsuccessful. The vinyl polyester sheets are quickly dried by blotting between layers of absorbent. Other surfaces can be dried by suspending them. A hair dryer can be used to shorten the drying time.

Kwik-Print is basically a contact-printing process requiring a negative transparency in order to produce a positive image. It is possible, however, to create an image via slide projection. In addition, thin paper negatives or black line drawings on acetate can be substituted for high-contrast film. Of course, photograms can also be made.

Prior to making the exposure, it is advisable to run an exposure test in half-minute increments. A Stouffer step tablet can be used to determine the best exposure for each color. A solid step 4 or step 5 usually indicates a good exposure time. An underexposed image will wash off completely during development. Overexposure is preferable, as some correction is still possible after "water development" by using the diluted ammonia solution or Kwik-Print Brightener.

To print multiple colors, it is only necessary to recoat the surface with pigment, re-expose, and redevelop the image. Prepunch all film prior to making tonal separations ensuring accurate registration of all color components when printing. Positive and negative transparencies may be combined, reversed, and/or printed separately. Lighter colors are usually printed first, followed by the darker pigments. Finally, where more than one copy of the image is being made, sensitized sheets (e.g., pigment applied and buffed dry) may be stored up to 20 hours prior to exposure in a lightproof container.

PHOTOGRAPHIC DRY-TRANSFER IMAGES

The Autotype Artsystem is a graphic-arts medium for making custom photographic dry transfers. The material is comparable to other preimaged dry-transfer sheets (e.g. stock alphabet sheets), in terms of transfer technique, permanency, and durability. Its major advantage is that any subject can be reproduced, making possible transfer of one's own photographic images to many surfaces: paper, plastic, metal, etc. Artfilm comes in the form of light sensitive sheets in sizes up to 51x61cm (20x24 in) and can be dyed in any of the seven Artsystem colors. These colors can also be mixed. To process Artfilm, it is first exposed to a transparency under ultraviolet light and then washed with water. It is next sprayed with Artsystem Dyestop, immersed in a dye bath, washed, and dried. Once dry, any segment of the image can be transferred to its final support by rubbing the back of the transfer material.

This transfer medium appears to offer many possibilities for image manipulation on transparent, as well as opaque, surfaces.

Intricate texture patterns reproduced photographically could be incorporated, in addition to more obvious imagery, for eventual selective transfer.

3M markets a light-sensitive dry-transfer material called Image n' Transfer (i.n.t.). The image is first created photographically by means of an intermediate negative transparency, or orange Color-Key, contact-printed with a sheet of i.n.t. material. Once the image has been developed, it can then be transferred to a receptor by the conventional dry-transfer burnishing technique. Unwanted image segments can be easily removed with sticky tape. Unlike Autotype Artsystem, 3M Image n' Transfer material is only furnished in black.

INSTANT ENLARGEMENTS WITH THE STABILIZATION PROCESS

There is commercial equipment available using the stabilization process for making enlarged continuous-tone reproductions from small negatives. One such piece of equipment is the Itek Reader Printer. It is a piece of equipment designed to make and deliver enlarged copies of documents that have been reduced onto micro-film. By placing a black-and-white or color negative transparency in the film holder, one can enlarge an image instantly to a print size dependent only on the unit's maximum capability: 46x61cm, 61x91cm, or 91x122cm (1½'x2', 2'x3', 3'x4'). After the light-sensitive paper in the reader-printer has been exposed, it is stabilized by chemical processing. This is all done internally within the machine itself and within a matter of seconds the print emerges. In the stabilization process the unexposed silver salts have not been removed in a hypo bath, however, the image will last for several years if it is not exposed to excessive humidity, heat, or direct sunlight. In order to make the image as permanent as any other photograph, immerse it in a normal print-fixing bath ("hypo") for about five minutes, then place it for another five minutes in a hypo-eliminator bath. Finally, wash the print for fifteen minutes in running water. The image can then be hand-colored or toned, for added effect. Unfortunately these units are expensive. Inquire of a local Itek dealer for companies willing to rent machine-time.

HANDPAINTED PROJECTIONS: A Manual Imaging System

Handpainted murals can be made via a projection system. The overhead projector lends itself to this application, as the unit is very simple to operate. A high-contrast transparency is placed on the glass imaging area, where it is magnified and projected. To make supersized graphics, a photographic image is first tonally separated into a series of high-contrast black-and-white transparencies. These image components are then projected, one at a time, onto a wall or other appropriate blank surface. (If the surface is outdoors projection will be more successful at night.) During projection, each image component is outlined by hand in pencil, to be filled in subsequently with a preselected color of paint. A code, perhaps of numbers, to identify each projected separation and its corresponding color of paint will prove helpful in the rendering of complex images. Each transparency must be carefully registered, outlined, and color-coded as it is projected. The remaining step in the process consists in manual completion of the image's color areas, using appropriate methods and materials. It is also possible to use a halftone image, rather than tonal separations, shading the dot pattern with the appropriate color. Coarse dot patterns are easier to trace, and the enlarged pattern does not necessarily hinder visual effectiveness. These units are available from audio-visual rental outlets and from 3M Company.

There are many old walls and playgrounds that could benefit from such manually executed images. Since the projected image will conform to the curves and bends in the surface, one is not limited to a flat area; but it is preferable that the surface have a light or white color for optimum visual contrast. An alternative to overhead projection is a normal slide projector, using positive litho transparencies. This approach is limited, however, by the size of the image that can be successfully projected from a small slide as well as by the wattage output of the projector's light source.

HOLOGRAPHIC IMAGERY

Holography is a new and rapidly growing visual technique capable of producing three-dimensional images of incredible accuracy and detail without the use of a camera lens. As one moves about a hologram, the image is seen from slightly altered perspectives, as it would be in reality. The effect is quite different from that of a flat, two-dimensional photograph. Special laser equipment is necessary in order to make holograms. The coherent light beam coming from the laser is split, one beam reflecting off the subject and setting up an interference pattern with the other component, the reference beam. Both beams meet at the film plane. This wave interaction is recorded in the film's emulsion. Strong light projected through the processed film reconstructs the image from the latent light-wave pattern stored on the film. This creates a very accurate visual replica of the original subject, as if suspended in midair. High-frequency sound waves have also been used to create acoustical holograms in a similar manner. In addition animated holograms are now feasible, as well as "inanimate" ones. Unfortunately, the cost of the necessary laser equipment and of an adequately rigged holographic studio has prohibited much experimentation of a nonscientific, noncommercial nature. For those interested in pursuing the subject, there are several sources listed in the appendix. Certainly, it is destined to become a major imaging system of the future.

KIRLIAN PHOTOGRAPHY

Kirlian photography is an "electrographic" process that employs electricity, rather than light, to create imagery. As with photograms, exposure is made by contact and no camera or lens is used. Objects are seen in silhouette and appear to be surrounded by radiance. Almost any object or subject can be investigated, such as stones, plants, fish, shells, etc., although a very dangerous shock or burn can result from using any human parts. In addition each electrically-produced image is recorded differently, revealing a multitude of colors and designs.

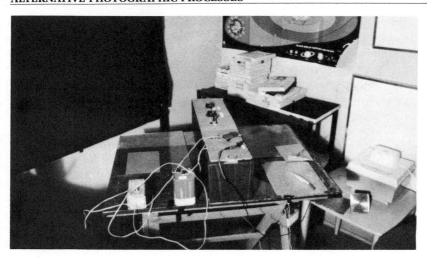

A Kirlian machine and materials.
Photo courtesy of Ann Chase.

Many individuals believe the coronalike patterns recorded on the film represent discharges of the life-energy or "The Force" itself. Others believe that these auras or flare patterns are the result of electrical impulses transmitted through the subject matter and that these transmissions are affected by the moisture content of the object. (The greater the amount of moisture contained within the object, the less its ability to transmit electrical energy, and less dramatic are the streamers recorded on the film.) Regardless of the various explanations, the beauty of these "light" emanations is apparent and worth exploring.

To produce a Kirlian image, a piece of film, either color or black and white, is placed, emulsion side up, on a metal plate. The object is placed on top of the film and sandwiched under another plate. Both plates are connected via electrodes to a low-wattage power source. (Kirlian kits are available, although all the necessary equipment can be assembled at home using a 6-volt battery.) The circuit is completed by switching on the generator. A pulsating, electrical current is delivered through the system, exposing the film and any object in contact with it. The film is then processed by conventional photographic darkroom techniques.

There are books that deal with the scientific and technical aspects of Kirlian photography. This brief outline is included simply to acknowledge another fascinating frontier in image making.

A Practical Conversion Table for work in small volumes

1 ounce = 28.35 grams
1 ounce = 437.5 grains
1 gram = 15.43 grains
1 gram = weight of 1cc water
1 U.S. nickel = 5 grams
1 pound = 16 oz.
1 pint = 16 fluid oz.
2 fl. oz. = 8 drams
1 fl. oz. = 29.57 cc
1 milliliter = 1 cubic centimeter
1 teaspoon = 5 ml or 5 cc
1 tablespoon = 15 ml or
   15 cc = 3 teaspoons = ½ oz.
1cc = 1ml =
   1/1000 of a liter = 15 drops
1 drop = 1 minim =
   approx. .067 ml
1 liter = 1000 cc = 33.81 fl. oz.
1 qt. = 0.95 liter = 32 fl. oz.
1 qt. = 2 pints
1 U.S. gallon = 4 qts
1 inch = 2.54 cm
1 cm = .39 inch
1 mm = .04 inch

Direct conversion formulas for both
minus and plus reading temperature

C=Celsuis F=Fahrenheit

$^{\circ}C=5/9(^{\circ}F-32)$.
$^{\circ}F=9/5(^{\circ}C+32)$.

Some common conversions:
50°F=10°C/68°F=20°C
95°F=35°C/122°F=50°C

Safe Handling of Chemicals
By Dr. Grant Haist

GENERAL PRECAUTIONS

The handling of chemicals for arts and crafts requires good laboratory practices if unfortunate consequences are to be avoided. The following general rules will help to minimize the dangers of working with chemicals whose potential harm to the artist should never be underestimated.

1. Treat all chemicals as if they were extremely dangerous, even though some may not be.

2. Keep containers closed except only when in use.

3. Use with adequate ventilation at all times.

4. Do not breathe vapors, avoid contact with skin or eyes, and do not swallow any compound or chemical preparation.

5. Wear protective clothing (rubber gloves, laboratory jackets or aprons, protect the eyes with safety glasses or shields, and wear a respirator.

6. Wash thoroughly with water after handling chemicals.

7. Do not smoke in the presence of chemicals or chemical preparations. Many are flammable or explosive.

8. Do not mix chemicals haphazardly, even during disposal.

9. If chemicals contact skin or eyes, or if swallowed, get medical attention at once.

10. Store chemicals in cool, dry areas away from sunlight.

11. Store away from children.

12. When diluting or dissolving strong acids or bases, always add the acid or base slowly to the surface of the water or liquid solution.

ACETIC ACID

This clear, colorless liquid has a very pungent vapor. Glacial acetic acid (99.5%) and photographic (28%) are two commonly available grades of this acid. Vinegar contains about 4 to 6% acetic acid. If stored below 60°F, glacial acetic acid will solidify but may be melted by placing the bottle in warm water.

Precautions: Acetic acid, especially the concentrated liquid (glacial acetic acid), causes severe burns. Use with adequate ventilation. Avoid contact with the concentrated liquid and do not breathe vapor. Keep glacial acetic acid away from excessive heat or open flame.

First Aid: In case of contact, use plenty of water to flush eyes and skin for about 15 minutes. Remove contaminated clothing and shoes. Wash clothing before use.

Use: To stop development of photographic silver-sensitized emulsions or to increase the intensity of the colors given by certain toners.

ACETONE

A colorless, volatile, flammable liquid that finds general use as an organic solvent.

Precautions: *Danger! Extremely flammable.* Keep container closed and away from heat, sparks or flame. Use adequate ventilation. Avoid long or repeated contact with the skin. Avoid contact with rayon clothing and stockings, plastics, and varnishes as acetone may dissolve these materials.

First Aid: If inhaled, remove to an area free of acetone fumes. Normal handling of acetone should not cause serious poisoning.

Use: As a solvent to solublize chemical compounds, and, diluted with an equal volume of water, as a means to soften the base of acetate photographic film to insure good contact for making exposures for photoetching.

ALCOHOL, DENATURED
This colorless, volatile, flammable liquid is usually ethyl alcohol to which an ingredient has been added to make the alcohol unsuitable for drinking. A variety of substances are used to denature ethyl alcohol, one of which is methyl alcohol (methanol). Methyl alcohol is harmful if inhaled, may cause blindness, and can be fatal. Therefore, treat denatured alcohol as if it contained a dangerous substance.

Precautions: *Danger! Flammable.* Keep container closed. Keep away from heat, sparks, and flame. Use with adequate ventilation.

First Aid: Methanol is a poison, and some of the other denaturing agents may also be dangerous. If inhaled, seek fresh air. If swallowed, give a tablespoon of Ipepac syrup with a glass of water to induce vomiting. Repeat one time only if no vomiting occurs within 20 minutes. If not breathing, give mouth-to-mouth artificial respiration.

Use: Denatured alcohol is used as a solubilizing agent for organic chemicals. It is also useful in removing unwanted portions of images that have been transferred from magazines or newspapers into acrylic polymer medium.

ALUMINUM POTASSIUM SULFATE (ALUM)
This colorless, odorless, crystalline solid is often called alum or potassium alum.

Precautions: No special precautions are necessary but safe laboratory practices in chemical handling and use should be followed.

Use: Used to harden the emulsion or remove the yellow stain in the gum bichromate process.

AMMONIA
This colorless and pungent gas forms ammonium hydroxide when dissolved in water (ammonia water).

Precautions: Do not mix ammonia with chlorine bleaches. Ammonium hydroxide can cause burns. Do not get in eyes, on skin, or on clothing. Avoid breathing the irritating vapor by using only in well-ventilated areas. Wash thoroughly after handling.

First Aid: In case of contact, use plenty of warm water to flush eyes or skin for about 15 minutes or more. Remove contaminated clothing and shoes. Wash clothing but discard contaminated shoes. If taken internally, give copious amounts of the juice of lemon, orange, or grapefruit, or diluted vinegar. Follow with olive oil. Get medical attention.

Use: An ingredient in pre-cleaning solution for metal plate surfaces, such as in photoetching. Ammonia fumes are used to develop diazo materials and for the correction of overexposure in the gum bichromate process.

AMMONIUM DICHROMATE (AMMONIUM BICHROMATE)
An odorless, orange-red, crystalline solid that is readily soluble in water. Ammonium dichromate is more sensitive to light than potassium dichromate.

Precautions: *Warning!* This compound is flammable. At about 225°C the decomposition of this compound becomes self-sustaining, with spectacular swelling, evolving heat, sparks or flame. Do not breathe dust or spray from solutions. Causes severe irritation, rash, or ulcers ("chrome sores").

First Aid: In case of contact, use plenty of water to flush skin for at least 15 minutes. If swallowed or in eyes, get medical attention quickly.

Use: Ammonium dichromate has many uses in the arts and crafts. A primary use is as a photoresist in the etching of metal, glass or ceramic surfaces. This compound is also used in the gum bichromate and mordant dye printing processes.

AMMONIUM PERSULFATE
Colorless crystals of a white powder that is readily soluble in water but decomposes in hot water.

Precautions: Ammomium persulfate is a strong oxidant. Fire may result from contact with other materials. Store in a cool, dry place, and away from combustible materials. Do not get in eyes, on skin, or on clothing. May cause skin irritation and is harmful if swallowed.

First Aid: Use plenty of water to flush skin if contact with the compound is made. Get medical help if compound is swallowed or in eyes.

Use: Helps to remove oxide from surface of metal pieces to be electroplated.

CARBOLIC ACID (PHENOL)
A white, crystalline, poisonous compound that is soluble in water.

Precautions: *Danger! May be fatal* if absorbed through the skin. Causes severe burns. Keep container closed. Avoid breathing vapor. Use only with adequate ventilation. Do not get in eyes, on skin, or on clothing. Wash thoroughly after handling.

First Aid: Poison. In case of any contact, use plenty of water immediately to flush skin and eyes for at least 15 minutes. Immediately remove contaminated clothing and shoes. Wash clothing thoroughly before using again but discard contaminated shoes. Get medical attention at once!

Use: Acts as a preservative for gum bichromate emulsions but alternative preservatives, possibly thymol, should be considered.

CAUSTIC SODA
(See Sodium Hydroxide)

CHROME ALUM
Chrome alum (potassium chromic sulfate dodecahydrate) is a violet, crystalline powder that is soluble in water, giving a violet solution in cool water and green in hot water.

Precautions: Chrome alum is a recognized carcinogen. This compound has a corrosive action on skin and mucous membranes of the nasal septum, causing deep, penetrating ulcers that are slow in healing and leave scars. Do not allow the compound or its solutions to contact the skin, eyes, and do not inhale dust or vapors. Chromium salts are known carcinogens of the nasal cavity, sinus, and lungs.

First Aid: In case of contact, flush skin or eyes with warm water for at least 15 minutes. Remove contaminated clothing and shoes. Get medical attention at once.

Use: Do not use! Substitute aluminum potassium sulfate (alum), a compound of much less toxicity.

COBALT CARBONATE

Red powder or crystalline compound that is almost insoluble in water or alcohol but soluble in dilute acid and ammonia.

Precautions: Harmful if swallowed. May cause skin irritation and allergenic reactions. Do not breathe dust. Keep container closed except when in use. Avoid prolonged and repeated contact with skin. Wash thoroughly after handling. Cobalt compounds are suspected carcinogens of connective tissue and lungs.

First Aid: If compound is contacted, flush skin and eyes with warm water for at least 15 minutes. Remove contaminated clothing and shoes. Wash clothing thoroughly before again using. Get medical attention.

Use: As a blue glazing material for forming photo images on ceramic materials.

COLLODION

A colorless or pale yellow, clear or slightly opalescent, viscous liquid that is a solution of pyroxylin in a 1:3 mixture of ethyl alcohol and ether.

Precautions: *Danger! Extremely flammable and explosive.* Keep container tightly closed except when in actual use. Keep away from heat, sparks or flame. Avoid breathing of vapor for any length of time. Use with adequate ventilation. Store in cool place.

First Aid: Get medical attention if contact is made with the eyes or if swallowed.

Use: Used as an overcoat in making photo images on ceramic surfaces.

COPPER CARBONATE

This compound is available as a blue green powder or dark green crystals, essentially insoluble in water but soluble in ammonia or dilute acids.

Precautions: *Poison.* Keep container closed. Do not breathe dust. Avoid contact with skin, eyes, or respiratory tract. May cause skin irritations or allergenic reactions. Wash with water after handling. Harmful if swallowed.

First Aid: If taken internally and if conscious, give a tablespoon of Ipepac syrup with a small glass of water to induce vomiting. Repeat dosage one time only if

vomiting does not occur within 20 minutes. Get medical attention while keeping the individual warm and quiet.

Use: As an ingredient for forming a blackish-green glaze for photo images on ceramic surfaces.

COPPER CYANIDE

Copper cyanide (cuprous cyanide) is a creamy white powder of a very poisonous nature that is insoluble in water.

Precautions: *Danger! Poison.* Do not contact compound with acid, acid gases, or steam. May be fatal if swallowed. Avoid contact with eyes and skin. Irritation of the skin ("Cyanide rash") is characterized by itching, eruptions, and often infection. Wash thoroughly after handling. Keep bottle closed. Store in a cool, dry place and away from acids. Acid may liberate poisonous gas. Do not breathe gas or dust. Hydrocyanic acid is a flammable gas of high toxicity.

First Aid: In case of contact with eyes, flush with plenty of water for at least 15 minutes. Remove contaminated clothing and remove to fresh air area. If gas is inhaled, break an amyl nitrate pearl in a cloth and hold lightly under nose for 15 seconds. Repeat 5 times at 15 second intervals. Use artificial respiration if breathing has stopped. If swallowed, repeat treatment with amyl nitrate pearl outlined above. If patient is conscious, give a tablespoon of Ipepac syrup with a small glass of water to induce vomiting. Repeat dosage one time only if vomiting does not occur within 20 minutes. Repeat inhalation of amyl nitrate five times at 15 second intervals. Use artificial respiration if breathing has stopped. In all cases, get medical attention.

Use: As a constituent of a plating bath for electroplating metals.

COPPER NITRATE

Large, blue-green, deliquescent crystals that are readily soluble in water. The blue, deliquescent crystals of the hexahydrate may lose three molecules of the water of hydration at about 26°C.

Precautions: *Poison!* May be fatal if swallowed. Keep container tightly closed. Avoid contact with skin, eyes and mucous membranes as copper salts are strong irritants. Do not breathe dust or solution spray. Do not swallow. Wash thoroughly after handling.

First Aid: If taken internally and if conscious, induce vomiting by giving a tablespoonful of Ipepac syrup in a small glass of water. Repeat this dosage one time only if no vomiting occurs within 20 minutes. Keep patient warm and quiet. Get medical attention at once.

Use: As a green colorant for decorating a photo-etched plate.

COPPER SULFATE

Cupric sulfate with 5 molecules of water is available as large, blue crystals, blue granules or a light blue powder. This compound is very soluble in water but only slightly soluble in ethyl alcohol.

Precautions: *Poison! May be fatal* if swallowed. Keep container closed. Do not breathe dust or solution spray. Avoid contact with skin, eyes and mucous membranes as copper salts are strong irritants. Do not swallow. Wash thoroughly after using.

First Aid: If taken internally and if conscious, induce vomiting by giving a tablespoon of Ipepac in a small glass of warm water. Repeat dosage one time only if vomiting does not occur within 20 minutes. Keep patient warm and quiet. Get medical attention at once.

Use: As a brown colorant for decorating a photoetched plate.

FERRIC CHLORIDE

This compound is available either as a brown-black, water-soluble solid (anhydrous form) or as orange-yellow crystals that are soluble in water. Ferric chloride is decomposed by air or moisture. It will take up moisture from the air and liquefy.

Precautions: Keep container closed except when in actual use. Avoid prolonged contact with skin and do not breathe vapor. Anhydrous form is an irritant and astringent. The hydrated form may give off hydrochloric acid. Compound is harmful if swallowed.

First Aid: Flush skin or eyes with plenty of warm water if contact is made. If swallowed, get medical attention.

Use: As an etchant in the resist-free areas of metals or as a hardener for a colloid, such as gelatin.

FORMALDEHYDE

Formaldehyde is a colorless, toxic gas with a suffocating odor that dissolves in water, forming formalin (about 35 to 40% formaldehyde). The vapors from formalin attack the eyes and respiratory system, causing intense irritation. Formalin may contain considerable amounts of an inhibitor, such as methyl alcohol, a poison.

Precautions: Keep container closed. Always use with plenty of ventilation. Avoid having formaldehyde come in contact with the skin, eyes, or clothing. Wash thoroughly after handling.

First Aid: In case of contact, immediately use plenty of water to flush eyes or skin for about 15 minutes. If swallowed, induce vomiting by giving a tablespoon of Ipepac syrup in a glass of water. Repeat this dosage one time only if vomiting does not occur within 20 minutes. If unconscious, do not give anything by mouth. Get medical attention.

Use: As a hardener for gelatin sizing of paper.

HYDROCHLORIC ACID

The water solution of this poisonous gas produces a clear, colorless, or slightly yellowish, fuming liquid with an irritating vapor.

Precautions: Hydrochloric acid causes severe burns. Keep container closed except only when in use. Use adequate ventilation. Avoid breathing vapor. Do not get in eyes, on skin, or on clothing. Wash with water after handling.

First Aid: In case of contact, use plenty of warm water to flush eyes or skin for about 15 minutes. Remove contaminated clothing and shoes. Wash clothing before using again. Get medical attention.

Use: To clean metal surfaces, as a means to etch metal or stone, or for deepening the blue coloration given by the cyanotype process.

HYDROFLUORIC ACID

The water solution of this colorless gas is extremely irritating to the skin.

Precautions: *Danger!* Causes extremely severe burns which may not be immediately visible or painful. May cause death if swallowed. Keep container closed except when in actual use. Do not breathe vapors. Do not get on skin, in eyes, or on clothing. Discard contaminated clothing. ALWAYS wear protective clothing, rubber gloves (two pair, preferably), and face shield. Store away from heat and out of direct sunlight.

First Aid: Always have available magnesium oxide-glycerin paste. If acid contacts skin, flush with cold water, particularly under fingernails, then apply paste. If acid contacts eyes, flush with plenty of cool water and get prompt medical attention. If swallowed, give water, followed by milk or milk of magnesia. Do not induce vomiting. Treat for shock. In any case of known or suspected contact, get immediate medical attention. Quick action is essential.

Use: Acid is an excellent etchant for glass but the dangers involved in the use of hydrofluoric acid should limit its application.

HYDROGEN PEROXIDE

A colorless, unstable liquid that mixes readily with water. The commercial grade contains about 30% hydrogen peroxide by weight.

Precautions: Hydrogen peroxide is a strong oxidant, and may cause burns and eye injuries. The effects may be delayed. Avoid contact with combustible materials. Do not contaminate hydrogen peroxide with metals, dust, or organic materials, because rapid decomposition may evolve large quantities of oxygen, generating high pressures. Use water only to treat fires involving hydrogen peroxide. Apply water quickly to entire spillage to flush away.

Store original vented container in a dry place away from heat and sunlight. Never return unused peroxide to bottle; dilute unused portion with water and discard. Rinse empty container with water before discarding or saving. Protect eyes with goggles or face shield. Wear neoprene, butyl rubber or vinyl gloves on hands.

First Aid: In case of contact, flush skin or eyes with

water for at least 15 minutes. Get medical attention, especially for eye contact or swallowing. Promptly remove contaminated clothing and shoes. Wash contaminated clothing thoroughly.

Use: As a means to deepen the blue coloration given by the cyanotype process or as a developer in the indirect method or applying an image to a silkscreen.

ISOPROPYL ALCOHOL
A colorless, flammable liquid soluble in water, having a weak alcohol odor.

Precautions: *Warning!* Isopropyl alcohol is flammable. Keep container closed. Use with adequate ventilation. Keep away from heat, sparks and flame.

First Aid: Inhalation of large quantities of vapor or swallowing isopropyl alcohol can cause dizziness, nausea, narcosis, and coma. Ingestion of 100 ml may be fatal. Get medical attention if ingestion or inhalation of large quantities is known to have occurred.

Use: Removal of pigment for image making on ceramic objects.

LIVER OF SULFUR (POTASH SULFURATED)
This substance is a mixture of potassium thiosulfate and potassium polusulfides. Liver of sulfur is available as yellowish-brown lumps which have a liver-brown color when freshly fractured. The water soluble lumps have a faint odor of hydrogen sulfide (rotten eggs) and decompose on exposure to air.

Precautions: Incompatible with acids, acid salts, carbonated waters, and alcohol. Keep container tightly closed. Hydrogen sulfide is a colorless, flammable, extemely harmful gas. Avoid inhaling gas or vapor from liver of sulfur solutions. Avoid contact with skin, eyes, or breathing vapor, as irritation may result. Wash thoroughly after handling.

First Aid: Loss of consciousness may result if much hydrogen sulfide is inhaled; remove patient to area of clear air. Keep warm and rest. Apply artificial respiration if breathing has stopped. Flush skin and eyes with water for about 15 minutes if contact is made. In severe cases, get medical attention.

Use: For darkening silver, copper and other metals.

NITRIC ACID
A colorless, or yellowish, fuming corrosive liquid that is soluble in water.

Precautions: Nitric acid produces an irritating vapor and causes severe burns. Keep container closed except for immediate use. Use with adequate ventilation. Avoid breathing the vapor. Do not get in eyes, on skin, or on clothing. Wash thoroughly after handling. Do not spill. Nitric acid may give off a dangerous gas or cause fire.

First Aid: In case of contact, use plenty of water to flush the eyes or skin for at least 15 minutes. Remove

contaminated clothing and shoes. Wash clothing before using again but discard contaminated shoes. Get medical attention.

Use: For cleaning or etching metal surfaces.

PHENOL (See Carbolic Acid)

PHOSPHORIC ACID
Phosphoric acid (orthophosphoric acid) is available as a clear, odorless, syrupy liquid or unstable crystals. When suitably diluted with water, the acid has a pleasant acid taste (used as an acidulant in cola-type beverages).

Precautions: The hot, concentrated acid attacks porcelain or granite ware but may be stored in glass or stainless steel containers. The concentrated acid can cause irritation. Avoid contact with the skin, the eyes, and clothing.

First Aid: In case of contact, flush eyes or skin with plenty of water for about 15 minutes. Remove and wash contaminated clothing. Get medical attention.

Use: As a mild etchant in photoetching.

POTASSIUM CHLORATE
This compound is a white or colorless, crystalline solid that is water soluble.

Precautions: *Dangerous!* Shock may explode. When heated, may emit poisonous fumes and explode. Do not mix with flammable substances or an explosion may result. Do not contaminate with charcoal, sawdust, shellac, starch, sugar, sulfur, sulfuric acid, or any combustible substance. Store in a cool, ventilated place, away from the hazard of fire, and away from all nearby combustible materials. Avoid contact with the skin, eyes or clothing. Do not swallow—chlorate is a poison.

First Aid: If compound is contacted, flush skin or eyes for about 15 minutes with plenty of water. Remove contaminated clothing and shoes. Wash clothing before using again. If ingested, call a physician at once. Potassium chlorate may damage the blood, kidneys, or heart muscles.

Use: As an ingredient in a copper etching bath.

POTASSIUM CYANIDE
This extremely poisonous, white granular powder, having a faint odor of almonds, dissolves readily in water.

Precautions: Do not contact compound with acid or acid gases. Do not swallow. *Poison!* May be fatal. Avoid contact with eyes and skin. Irritation of the skin (cyanide rash) is characterized by itching, eruptions, and often infection. Wash thoroughly after handling. Keep bottle closed. Store in a cool, dry place, away from acids. Acid may liberate poisonous gas that is flammable. Do not breathe gas or dust. Do not take internally.

First Aid: In case of contact with eyes, flush with plenty of water for about 15 minutes or more. Remove contaminated clothing and take patient to a fresh air area. If gas is inhaled, break an amyl nitrate pearl in a cloth and hold lightly under nose for 15 seconds. Repeat 5 times at 15 second intervals. Use artificial respiration if breathing has stopped. If swallowed, repeat treatment with amyl nitrite pearl outlined just above. If patient is conscious induce vomiting by giving a tablespoon of Ipepac syrup in a glass of water. Repeat dosage one time only if vomiting does not occur in 20 minutes. Repeat inhalation of amyl nitrite five times at 15 second intervals. Use artificial respiration if breathing has stopped. In all cases, get medical attention quickly.

Use: As a constituent of a plating bath for electroplating metals. Use of this compound should be avoided, however.

POTASSIUM BICHROMATE or POTASSIUM DICHROMATE
This salt, referred to either as a bichromate or a dichromate, is a crystalline, orange-red compound that is readily soluble in water. When stuck by light, this compound renders gelatin and other suitable colloids insoluble in water.

Precautions: Dichromate is a poison and may cause cancer. Keep container closed. Avoid contact with skin, eyes, and clothing. Avoid breathing dust or solution spray. Dichromate causes skin irritation, rash, or ulcers, as well as destruction of mucous membranes. May cause cancer of the lung. Swallowing can result in death—thirty grams have been reported to be fatal.

First Aid: In case of contact, flush skin or eyes with water for at least 15 minutes. In case of any contact or swallowing, get medical attention at once.

Use: As a light sensitizer to harden suitable colloids, such as in the making of resists for etching metal, glass or ceramic surfaces.

POTASSIUM FERRICYANIDE
A bright red, crystalline, poisonous solid that is soluble in water.

Precautions: Harmful if swallowed. Toxic fumes are released when heated or in contact with acids. Keep container closed and away from heat and acids. Protect solution from light. Wash thoroughly after handling.

First Aid: Antidote: If taken internally, promptly give a tablespoon of dilute hydrogen peroxide (3%). Apply cold water to head and spine. Give inhalation of ammonia and, if necessary, artificial respiration.

Use: As an etchant for silver bromide prints or as part of the sensitizer in the cyanotype (blueprint) process.

POTASSIUM PERMANGANATE
This powerful oxidizing agent is a dark purple, crystalline solid that is soluble in water. Stains given by this compound may be removed by a solution of sodium bisulfite or sodium metabisulfite.

Precautions: Do not contact potassium permanganate with glycerine or ethylene glycol. *Danger!* Strong Oxidant. Do not store near, or use the compound with, combustible materials. Keep container closed. Can cause skin irritation. Harmful if swallowed.

First Aid: Do not get on skin, in eyes or on clothing. Wash thoroughly after use. If taken internally, give large amounts of water or strong tea. Give an emetic (mustard).

Use: As a reducing agent for the silver image on a photographic print.

SELENIUM
This element exists in several forms, often in a dark red crystal or a red liquid compound.

Precautions: Selenium compounds are toxic. Avoid having selenium compounds contacting the skin or eyes, and do not breathe dust or fumes. May cause skin irritation or irritation of the respiratory tract. Garlic odor of breath is a symptom. Keep container closed. Wash thoroughly after using.

First Aid: In case of contact with skin or eyes, flush with plenty of water for at least 15 minutes. Do not swallow as selenium compounds are suspected carcinogens of the liver and thyroid. If swallowed, get medical attention at once.

Use: As a toning bath for converting silver photographic images into pleasant reddish tones.

SILVER NITRATE
An odorless, colorless or white, crystalline, poisonous compound whose water solutions stain human skin. The pure compound is not darkened by light but darkening occurs quickly if contaminated by organic matter or hydrogen sulfide.

Precautions: *Poison!* Incompatible with a considerable number of compounds so do not contact with other compounds unless known to be safe. Causes burns. Stains skin. Do not get on skin, in eyes, or on clothing. Irritating to skin and mucous membranes. Swallowing can cause death. Keep container closed. Do not contaminate clothing and protect the face and eyes. Wash thoroughly after using.

First Aid: In case of contact with eyes, immediately flush with water for at least 15 minutes. If taken internally, induce vomiting by giving one tablespoon of Ipepac syrup in a small glass of water. Repeat the dosage one time only if no vomiting occurs within 20 minutes. Have the patient lay down and keep warm. Get medical attention at once.

Use: As an ingredient in a photographic sensitizer for paper or cloth.

SODIUM HYDROXIDE (CAUSTIC SODA)
This white, water-soluble solid (available as granules,

pellets, lumps, or sticks) generates considerable heat when dissolved in water. When making solutions, add the solid hydroxide slowly to the surface of the solution to avoid spattering.

Precautions: Caustic soda causes severe burns to skin and eyes, and may be fatal if swallowed. Avoid contact with eyes, skin, or clothing. Do not take internally. Protect eyes with glasses, goggles, or face shield. Keep container closed except when in actual use. Caustic soda absorbs moisture and carbon dioxide from air, reducing its alkalinity.

First Aid: If contacted, immediately flush skin or eyes with plenty of water for at least 15 minutes. For eyes, get medical attention. If taken internally, give water containing large amounts of diluted vinegar, or orange, lemon, or grapefruit juice. Follow with milk or white of eggs beaten with water. Get medical attention at once.

Use: As an etchant for linoleum.

SULFURIC ACID

A clear, colorless, odorless, oily liquid that dissolves in water or alcohol with the evolution of much heat and with a contraction of volume. Because of its great attraction for water, sulfuric acid removes water from the air or from organic substances, causing them to char in the process.

Precautions: *Danger!* Causes severe burns. Keep container closed and handle with caution. Sulfuric acid is corrosive to all body tissues, so do not allow contact with skin, eyes, or clothing. Do not inhale fumes. Ingestion may cause death. Always add acid slowly to water when diluting the acid. Wash thoroughly after handling.

First Aid: In case of contact, immediately flush eyes or skin with water for at least 15 minutes. Remove contaminated clothing and shoes. Wash clothing before re-use. Get immediate medical attention.

Use: As an ingredient in bleaches for silver images (photographic).

TRICHLOROETHYLENE

This colorless, nonflammable liquid has an odor characteristic of chloroform. Trichloroethylene is practically insoluble in water but soluble in alcohol. The compound may slowly decompose in the presence of of light and moisture.

Precautions: Harmful if inhaled. Keep container closed. Use with adequate ventilation. Avoid breathing

vapor. Moderate exposure may cause inebriation, increased exposure may have a narcotic effect, and heavy exposure may cause death. Avoid prolonged or repeated contact with the skin.

First Aid: If inhaled, remove to fresh air. If not breathing, give mouth-to-mouth respiration. Give oxygen if breathing is difficult. Get medical attention.

Use: As a developer for presensitized plates.

TRISODIUM PHOSPHATE

Sodium phosphate, tribasic, is a colorless or white crystalline compound with 12 molecules of water, forming strongly alkaline water solutions.

Precautions: Keep container closed. Avoid contact with skin, eyes, or clothing. Can cause skin irritation. Harmful if swallowed. Wash after using.

First Aid: If contacted, flush skin or eyes with water for about 15 minutes. If swallowed, get medical attention.

Use: As a degreaser in the preparation of silkscreens.

VINEGAR

Vinegar is a weak solution of impure acetic acid, commonly 4 to 5% acetic acid as sold in grocery stores.

Precautions: Use with adequate ventilation. Avoid contact with skin, eyes or clothing. Do not breathe vapor. Keep container closed.

First Aid: In case of contact, use plenty of water to flush eyes and skin for about 15 minutes. Remove contaminated clothing and shoes.

Use: To help clean silkscreen mesh.

XYLENE

An oily, colorless, toxic, flammable liquid that is practically insoluble in water but soluble in absolute alcohol and other organic solvents.

Precautions: *Danger! Flammable.* Keep container closed and away from heat, sparks, or flame. Use adequate ventilation. Avoid long or repeated contact with skin.

First Aid: Avoid breathing as may be narcotic in high concentrations. Human toxicity is not well defined but is less toxic than benzene. Avoid contact with skin, eyes, or clothing, and do not swallow. Get medical attention. In case of fire, use foam, carbon dioxide, or dry chemicals.

Use: To help clean silkscreen after using oil-based inks.

Source List of Suppliers, Tools, Equipment

METAL APPLICATIONS
Industrial Photographic System Sales
Eastman Kodak Company
343 State Street
Rochester, NY 14650
or your local Kodak dealer
Photo Resists.

Questions regarding safe handling of
Photo Resists should be directed to:
Laboratory of Industrial Medicine
Eastman Kodak Company
343 State Street
Rochester, NY 14650

Support literature available on Kodak Photo Resists is
listed in the *Annual Index to Kodak
Information*—write for free copy a:
Dept. 412-C
Eastman Kodak Company
343 State Street
Rochester, NY 14650

Dynachem Corporation
1670 S. Amphlett Blvd.
San Mateo, CA 94402
Photo Resists.

Graphic Chemical & Ink Co.
728 N. Yale
Villa Park, IL 60181
*Printmaking supplies & equip., Litho equip., etching
tools for woodblock & metal, solvents, inks.*

Norland Products, Inc.
695 Joyce Kilmer Ave.
New Brunswick, NJ 08902
Photo Resists.

United American Metals Co.
2448 E. 25th St.
Los Angeles, CA 90058
Presensitized 16 gauge magnesium.

Thomas C. Thompson Co.
1539 Old Deerfield Road
Highland Park, IL 60035
Enamels for metal plus accessories.

The Ceramic Coating Co.
P.O. Box 370
Newport, KY 41072
Enamels for metal including aluminum applications.

GLASS APPLICATIONS
Atlas Screen Printing Supplies
1733 Milwaukee Ave.
Chicago, IL 60647
Low fired glass epoxy inks.

Bullseye Glass Co.
3722 S.E. 21st Ave.
Portland, OR 97202
Stain glass manufacturer.

Genesis Glass, Ltd.
P.O. Box 12307
Portland, OR 97212
Stain glass manufacturer.

Gloucester Co., Inc.
235 Cottage St.
Franklin, MA 02038
Phenoseal: a vinyl based bonding adhesive.

McKay Chemical Co.
880 Pacific St.

Brooklyn, NY 11238
Screen Etch Compound for glass.

Naz-Dar Company
1087 N. North Branch St.
Chicago, IL 60622
Epoxy resin inks for use on glass surfaces.

Norland Products, Inc.
695 Joyce Kilmer Ave.
New Brunswick, NJ 08902
Glue/silver chloride emulsion, NGS II (silver images on glass).

L. Reusche & Co.
2 Lister Ave.
Newark, NJ 07105
Glass color pigments, firing tech. for glass decals, matte etchants.

Rockland Colloid Corp.
500 River Road
Piermont, NY 10968
Print-E-Mulsion series for printing on paper and other materials.

Standard Ceramic Supply
Division of Chem-Clay Corp.
P.O. Box 4435
Pittsburgh, PA 15205
Glass colors.

CERAMIC APPLICATIONS

AMACO (American Art Clay Co.)
4717 W. 16th
Indianapolis, IN 46222
Versa colors.

Atlas Screen Printing Supplies
1733 Milwaukee Ave.
Chicago, IL 60647
Simplex decal paper, covercoat varnish, grinding vehicles.

H. Battjes
5507 20th W.
Bradenton, FL 33507
Custom one-color ceramic decals from own images.

Brittains (U.S.A.) Ltd.
26 Strawberry Hill Ave.
Stamford, CT 06902
Thermaflat decal paper.

Cerami-Corner, Inc.
Box 516
Azusa, CA 91702
Premade multi-color decals of various subject matter.

The Forming Company
1400 N. W. Kearney St.
Portland, OR 97209
Ceramic supplies & equip., stain glass, enamels.

Leslie Ceramic Supply Co.
1212 San Pablo Ave.
Berkeley, CA 94706
China paints, lusters, decals, transparent glaze base, metallic oxides, tiles, kilns.

Naz-Dar Company
1087 N. North Branch St.
Chicago, IL 60622
Encosol-1, 2, and 3.

Richards Pug Mill
8065 S.E. 13th Ave.
Portland, OR 97202
Pottery supplies and equipment.

L. Reusche & Co.
2 Lister Ave.
Newark, NJ 07105
Overglaze colors, printing mediums, overcoat varnish, and liquid lusters for ceramic decals.

Rockland Colloid Corp.
500 River Road
Piermont, NY 10968
Print-E-Mulsion series for printing photographic images on paper and other surfaces.

Standard Ceramics Supply Co.
P.O. Box 4435
Pittsburgh, PA 15205
Overglaze colors & mediums, clays, chemicals, kilns, pottery tools.

Van Howe Ceramic Supply Co.
11975 East 40th Ave.
Denver, CO 80239
Commercial decals.

FABRIC APPLICATIONS

Advance Process Supply Co.
400 N. Noble St.
Chicago, IL 60622
Transcello 4-color process photo silkscreen subliminal inks (Hit Series).

Air Photo Supply Corp.
158 South Station
Yonkers, NY 10705
Luminos photo linen.

Calcom Graphic Supply
2836 10th Street
Berkeley, CA 94710
Watertex fabric inks, photographic screen printing supplies.

Cerulean Blue, Ltd.
1314 N. E. 43rd St.
Seattle, WA 98105
Dye, fabric & related chemicals.

Colonial Printing Ink Co.
Div. of U.S. Printing Ink Co.
180 E. Union Ave.
East Rutherford, NJ 070703
Colonial Trans-Fab sublimation process colors for screening & lithographic applications.

D.Y.E. Textile Resources
3763 Durango Ave.
Los Angeles, CA 90034
Dyes, fabrics, blue print & brown print sensitizer kits.

Rockland Colloid Corp.
500 River Road
Piermont, NY 10968
Fabric sensitizer (FA-1), contact speed photo chemistry.

Screen Process Supply Co.
1199 E. 12th St.
Oakland, CA 94606
Inko dye, registration paste.

Testfabric, Inc.
200 Blackford Ave.
Middlesex, NJ 08846
Fabric source.

Zellerbach Paper Co.
9111 N. E. Columbia Blvd.
Portland, OR 97220
Transeal heat transfer paper (300-J), Patina coated matte, and Z-Heat transfer paper.

"OTHER" SURFACES
Hunt Mfg. Co.
1405 Locust St.
Philadelphia, PA 19102
Woodprint tools, inks, linoleum blocks, etc.

Consolidated Stamp Co.
589 Mission St.
San Francisco, CA 94105
Bakelite, Vulcanizers, etc.

L. Reusche & Co.
2 Lister Ave.
Newark, NJ 07105
Glass etchant for rubber stamps.

Naz-Dar Company
1087 N. North Branch St.
Chicago, IL 60622
Complete line of inks for screen printing on plastics.

Rockland Colloid Corp.
500 River Road
Piermont, NY 10968
Projection speed paint-on photo emulsions.

ALTERING THE SILVER IMAGE
Artistic Mfg. Co.

Division of Sun Chemical Corp.
Carlstadt, NJ 07072
Lift-A-Print materials.

Delta Import
319 West Erie St.
Chicago, IL 60610
Argenta monochrome color paper.

Luminos Products
% Air Photo Supply Corp.
158 South Station
Yonkers, NY 10705
Luminos monochrome color paper.

John G. Marshall Mfg., Inc.
167 N. 9th St.
Brooklyn, NY 11211
Marshall photo oils, protective lacquer sprays, toothing lacquers.

McDonald Photo Products, Inc.
2522 Butler
Dallas, TX 75235
Retouching lacquer, (Quebrada) decoupage information.

Peerless Color Laboratories
11 Diamond Place
Rochester, NY 14609
Transparent photographic watercolors.

Photo Technical Products Group
623 Stewart Ave.
Garden City, NY 11430
Ademco heat seal films, dry mounting, laminating equip., and accessories.

Sangray Corp.
Box 2388
Pueblo, CO 81004
Decalon, the instant decal medium; Decalcomania booklet.

FILMS, CHEMICALS, FRAMES, TRADITIONAL PHOTOGRAPHIC PROCESSES
A.S.F. Sales
P.O. Box 6026
Toledo, OH 43614
Quality metal frames. Mail order—toll free phone.

Elegant Images
2637 Majestic Drive
Wilmington, DE 19810
Paper, chemicals, introductory packets for Platinum and Palladium printing.

Free Style Sales Co., Inc.
5124 Sunset Blvd.
Los Angeles, CA 90027
Small quantity packaging or large sheets Ortho film. Mail order.

Gallery 614, Inc.
614 W. Berry St.
Fort Wayne, IN 46222
Materials & instructions for carbon and carbo printing, monochrome pigmented tissues, Tri-color Carbo tissue, transfer papers, chemicals, non supercoated silver bromide paper.

Nurnberg Scientific
6310 S. W. Virginia Ave.
Portland, OR 97201
Chemicals & laboratory supplies.

Treck Photographic
1919 S. E. Belmont
Portland, OR 97214
or offices in major cities
Photographic supply house: photo resists, graphic arts film, photographic chemicals, photo oils, Premier contact printing frames, PERFEX photographic opaque, etc.

PHOTO SILKSCREEN SUPPLIES (General)
Advance Process Supply
400 N. Noble St.
Chicago, IL 60622

Atlas Silkscreen Supply
1733 Milwaukee Ave.
Chicago, IL 60647

Naz-Dar Company
1087 N. North Branch St.
Chicago, IL 60622
Screen printing equipment, films, emulsions, etc.

McGraw Colorgraph Co.
175 West Verdugo
Burbank, CA 91503
Photo stencil film, carbon tissue, photogravure resist materials.

Ulano Photosilkscreen Products
210 E. 86th St.
New York, NY 10028

GRAPHIC COMMUNICATIONS EQUIPMENT AND SUPPLIES
AZO Plate
Murray Hill, NJ 07974
Enco Naps and Paps color proofing film.

AutoType, U.S.A.
501 West Golf Road
Arlington Heights, IL 60005
Photographic dry transfer materials and dyes.

California Ink Co.
501 15th St.
San Francisco, CA 94103
Inks, chemicals, and litho plates.

Direct Reproduction Corp.
835 Union St.
Brooklyn, NY 11215
Kwik-Print Materials.
Distributed by
Light Impressions Corp.
131 Gould Street
Rochester, NY 14610

E. I. du Pont de Nemours & Company
Photo Products Dept.
Wilmington, DE 19898
Cromalin color proofing system.

Itek Business Products
Div. of Itek Corp.
Rochester, NY 14603
Itek Reader Printer.

Lith-Kem Co.
46 Harriet Place
Lynbrook, L.I., NY 11563
Aluminum plates.

Polychrome Corp.
P.O. Box 817
Yonkers, NY 10702
or local branch
Polychrome Wipon sensitizing solution (909/910) and plate developer (305) kit.

3M Company
Printing Products Div.
3M Center
St. Paul, MN 55101
3M color key, and transfer key materials. Grained presensitized litho plates.

Western Litho Plate
3433 Tree Court
Industrial Blvd.
St. Louis, MO 63122
Ball grain aluminum plates.

Xerox Corporation
Color Copier Div.
Xerox Square
Rochester, NY 14644
Xerox 6500 Color Copier.

"OTHER" IMAGING SYSTEMS
Dynell Electronics Corp.
Marketing Dept.
Mexville, NY 14600
Solid photographic systems: Sculpt in bronze, wax, or other materials from 3-D originals.

Edmund Scientific Co.
300 Edscorp Bldg.
Barrington, NJ 08007
Kirlian electrophotographic kits, specialized photographic accessories/catalog no. 761.

Holosonics, Inc.
Richland, WA 98105
Holography.

New York School of Holography
120 W. 20th St.
New York, NY 10011

Museum of Holography
11 Mercer St.
New York, NY 10011
A gallery.

Multiplex Co.
454 Shotwell St.
San Francisco, CA 94105
Custom made 360° animated holograms.

The Whole Image
3609 S. E. Division
Portland, OR 97202
Custom Holograms.

FILMS AND SLIDES

Manson & Kennedy
2111 S. W. Vista Ave.
Portland, OR 97201
Films on printmaking technique. Cover photoetching, aquatinting, color printing, embossing, etc.

New Photographics Slide Bank
% Central Washington State College
Art Department
Ellensburg, WA 98926
Extensive collection of slides available to teachers and schools. Covers alternative photographic processes, also conventional imagery.

Bibliography

These references have been classified according to the areas for which I found them most useful. Therefore, they are not necessarily classified as their authors may have classified them, nor are they classified as to all of the subject matter contained within them, and the dates are those of the editions I am using. In addition, many of them have overlapping areas of interest relevant to this book.

*Indicates a catalog or pamphlet

GENERAL SOURCES, ENCYCLOPEDIAS AND REFERENCES FOR IMAGE IMPROVEMENT

Adams, Ansel, **The Basic Photo Series**, Books 1-5. Hastings-on-Hudson, New York, Morgan & Morgan, 1968.

Boyd, Jr. Harry, **A Creative Approach to Controlling Photography**. Austin, Texas, Heidelberg Publishers, 1974.

Boucher, Paul E., **Fundamentals of Photography**. New York, D. Van Nostrand Company, 1940.

Cassell's Cyclopaedia of Photography, Bernard E. Jones, ed. New York, Arno Press, 1973. A facsimile edition of the classic.

Creative Darkroom Techniques. Rochester, New York, Eastman Kodak Company, 1973.

Dowdell III, John J., and Zakia, Richard D., **Zone Systemizer for Creative Photographic Control**. Dobbs Ferry, New York, Morgan & Morgan, 1973.

Feininger, Andreas, **Principles of Composition in Photography**. Englewood Cliffs, New Jersey, Prentice-Hall, 1973.

Feininger, Andreas, **Total Picture Control**. Englewood Cliffs, New Jersey, Prentice-hall, 1970.

Focal Dictionary of Photographic Technologies, The, D. A. Spencer, ed. London, Focal Press, Ltd., 1973.

Focal Encyclopedia of Photography, The. London, Focal Press, Ltd., 1965.

Frontiers of Photography. New York, Time-Life Books. 1972.

History and Handbook of Photography, A, Translated From the French of Gaston Tissandier, Thomson, J., ed. New York, Arno Press, 1973. A facsimile reprint of the classic.

Kelley, Kathleen, **Testimony to a Process: Variations of Photographic Printing and Projection Processes**. Portland, Oregon, Portland Art Museum, 1974.

Kodak Reference Handbook: Materials, Processes, Techniques. Rochester, New York, Eastman Kodak Company, 1951.

Mante, Harold, **Photo Design: Picture Composition for Black and White Photography**. New York, Van Nostrand Reinhold Co., 1971.

Merck Index: An Encyclopedia of Chemicals and Drugs, The, Paul G. Stecker, ed. Rahway, New Jersey, Merck & Co., Inc., 1968. Contains definitions of all chemicals and their usage.

Neblette, C. B., **Photography Principles and Practices**. New York, D. Van Nostrand Company, 1927.

Petersen's Guide to Creative Darkroom Techniques. Los Angeles, Petersen Publishing Co., 1973.

Photo-Lab-Index: The Cumulative Formulary of Standard Recommended Photographic Procedures, Ernest M. Pittaro, ed. Dobbs Ferry, New York, Morgan & Morgan, 1977.

Picker, Fred, **The Fine Print**. Garden City, New York, Amphoto, 1975.

Picker, Fred, **Zone VI Workshop**. Garden City, New York, Amphoto, 1974.

Pictorial Cyclopedia of Photography, The. Cranbury, New Jersey, A. S. Barnes & Co., 1968

Print, The. New York, Time-Life Books, 1970.

Swedlund, Charles, **Photography: A Handbook of History, Materials, and Processes**. New York, Holt, Rinehart and Winston, 1974.

Stroebel, Leslie, and Todd, Hollis N., **Dictionary of Contemporary Photography**. Dobbs Ferry, New York, Morgan & Morgan, 1974.

Vogel, Herman, **The Chemistry of Light and Photography**. New York, Arno Press, 1973. A facsimile of the classic.

Wheeler, Capt. Owen, **Photographic Printing Processes**. London, Chapman & Hall, Ltd., 1930.

White, M., Zakia, R., and Lorenz, P., **The New Zone System Manual**. Dobbs Ferry, New York, Morgan & Morgan, 1976.

Woell, Fred F., **Photography in the Crafts.** Deer Isle, Maine, privately published, 1974.

Zakia, Richard D., **Perception and Photography**. Englewood Cliffs, New Jersey, Prentice-Hall, 1975.

PHOTO PRESENTATION ON METAL SURFACES
A. Electroplating
Electroplating. New York, Popular Science Publishing Co., 1936.

Untracht, Oppi, **Metal Techniques for Craftsmen**. New York, Doubleday & Company, 1968.

Metal Finish Guidebook and Directory. Westwood, New Jersey, Metal and Plastics Publications, 1971.

Plating Made Easy, Bulletin 347. Buffalo, New York, Hoover and Strong, Inc., 1947.

B. Enameling Metal
Ball, Fred, **Experimental Techniques in Enameling**. New York, Van Nostrand Reinhold Company, 1972.

Rothenberg, Polly, "*Champlevé Enamels,*" **Ceramics Monthly**, November 1966, pp. 28-29.

Seeler, Margaret, **The Art of Enameling: How to Shape Precious Metal and Decorate It with Cloisonné, Champlevé, Piquè-à-Jour, Mercury Gilding and Other Fine Techniques**. New York, Van Nostrand Reinhold Company, 1969.

Untracht, Oppi, **Enameling on Metal**. New York, Greenberg Publisher, 1957.

*Amaco Catalog: **Pottery and Metal Enameling Supplies and Equipment**. American Art Clay Co., Inc., 4717 West 16th Indianapolis, Indiana 46222.

C. Jewelry Making
Morton, Philip, **Contemporary Jewelry: A Studio Handbook**. New York, Holt, Rinehart and Winston, 1970.

Moty, Eleanor, "*Workshop: Photofabrication*," **Craft Horizons**, June 1971, pp. 14-17.

Penland School of Crafts: Book of Jewelry Making, The, John Payne, ed. New York, Rutledge Books, 1975. Chapter "*Mirror Image,*" by Eleanor Moty.

Von Neumann, Robert, **The Design and Creation of Jewelry**. Radnor, Pennsylvania, Chilton Book Company, 1972.

Tools, Supplies, and Equipment for Technicians and Craftsmen: A Jeweler's Supply Catalog No. 755. Montana Assay Office, Portland, Oregon 97204.
(li space)

D. Photoetching: Technique and Materials
German, D.E.B., and Stewart, J.C.J., "*A Simplified Method for Making Accurately Etched Shapes in Molybdenum Foil by a Photo Sensitive Resist Technique,*" **Journal of Photographic Science**, 15, May-June 1967, pp. 137-144.

Hepher, M., "*The Photo-Resist Story: From Niépce to the Modern Polymer Chemist,*" **Journal of Photographic Science**, 12, July-August 1964, pp. 181-190.

Kosloff, Albert, **Screen Printing Electronic Circuits**. Cincinnati, Signs of the Times Publishing Co., 1968.

Basic Principles of Photo Chemical Machining, PCM1-G-100, Technology Profile. Evanston, Illinois, Photo Chemical Machining Institute, 1975.

*E.I. du Pont de Nemours & Co., Inc., Photo Products Dept., 3300 West Pacific Ave., Burbank, California. Write for literature on **Photopolymer Film Resist Systems**.

*Eastman Kodak Publications:
G-38 **Kodak Photosensitive Resist Products for Photofabrication**.
P-79 **An Introduction to Photofabrication Using Kodak Photosensitive Resists**.
P-81 **Kodak Photo Resist and Kodak Photosensitive Lacquer**.
P-82 **Kodak Metal Etch Resist**.
P-83 **Kodak Ortho Resist**.
P-86 **Kodak Photo Resist, Type 3**.
P-88 **Kodak Thin Film Resist**.
P-137 **Kodak Photo Resist, Type 4**.
P-173 **Selecting, Using and Troubleshooting Kodak Photo Sensitive Resists**.
P-215 **Nameplates by Photofabrication With Kodak Photosensitive Resists**.
P-246 **Photofabrication Methods With Kodak Photosensitive Resists**.
Write: Eastman Kodak Company, Dept. 454, Rochester, New York 14650.

*Norland Technical Data:
Norland Glue-Silver Emulsion, NGS11.
Norland Water Based Photo Resists.
Norland Photo Resist 6 (NPR 6).
Norland Production Procedures for Anodized Nameplates.
Norland Silkscreen Emulsion.
Chemical Milling Precision Parts.
"*Water Base Photo Resists,*" **Photo Chemical Machining—Photo Chemical Etching**.
Write: Norland Products, Inc., 695 Joyce Kilmer Ave., New Brunswick, New Jersey 08902.

*Thiokol/Dynachem Technical Data Sheets:

CMR 5000 Chemical Milling Photo Resists.
Dynachem Photo Resist Instruction Outline.
Why DCR Is the Leading Photo Resist.
Dynacure tm SR-25 UV (UltraViolet) Cure Screen Resist.
Write: Thiokol/Dynachem Corp., 1670 South Amphlett Blvd., Suite 203, San Mateo, California 94402.

INTAGLIO PRINTMAKING
Banister, Manly, **Etching and Other Intaglio Techniques**. New York, Sterling Publishing Co., 1969.
Chamberlain, Walter, **Etching and Engraving**. New York, The Viking Press, 1972.
Heller, Jules, **Printmaking Today: An Introduction to the Graphic Arts**. New York, Henry Holt and Co., 1958.
Ross, John, and Romano, Clare, **The Complete Printmaker**. New York, The Free Press, 1972.

PHOTO PRESENTATION ON GLASS SURFACES:
TECHNIQUES AND MATERIALS
Bernstein, Paul, "*Holography*," **Glass Art Magazine**, Vol. 5, No. 4, pp. 20-22. Article deals with holography and glass art.
Brezhnoi, Anatolii, **Glass-Ceramics and Photo Sitalls**. London, Plenum Press, 1970.
Campana, D.M., **Acid Etching on Porcelain and Glass**. Chicago, D.M. Campana Co., 1946 (privately published).
Gick, James E., **Creating With Stained Glass**. Laguna Hills, California, Future Crafts Today, 1976. Information on sandblasting and screen etch technique on pp. 42-43.
Isenberg, Anita dn Seymour, **How to Work in Stained Glass**. New York, Chilton Book Company, 1972.
Kinney, Kay, **Glass Craft: Designing, Forming, Decoration**. New York, Chilton Book Company, 1962.
Quagliarta, Narcissus, **Stained Glass From Mind to Light**. San Francisco, Mattole Press, 1976.
Rigan, Otto B., **New Glass**. San Francisco, San Francisco Book Co., 1976.
Schloes, S.R., **Modern Glass Practice**. Chicago, Industiral Publication, Inc., 1952.
*Eastman Kodak Publications:
P-140 **Characteristics of Kodak Photographic Plates.**
P-245 **Decorating Glass Using Kodak Photosensitive Resists.**
Write: Eastman Kodak Company, Department 454, 343 State Street, Rochester, New York 14650.

PHOTO CERAMICS: TECHNIQUE AND MATERIALS
Bartel, Marvin, "*Silk Screening With Slip*," **Ceramics Monthly**, March 1973, pp. 26-31.
Burbank, W.H., **Photographic Printing Methods**. New York, New York, Arno Press, 1973. A reprint of the 1891 edition. Chapter on photo ceramics.
Campana, D.M., **Photo Ceramics**. D.M. Campana Co.,

Chicago, Ill., 1945 (self-published). Recipes for ceramic photography to be burned into porcelain or glass.
Goldman, Caren, "*Picture Pots*," **Ceramics Monthly**, January 1968, pp. 20-21.
Kaplan, Jonathan, "*Making Ceramic Decals*," **Ceramics Monthly**, April (Part I) and May (Part II) 1975.
Kenny, John B., **Ceramic Design**. New York, Chilton Book Company, 1963.
Kosloff, Albert, **Ceramic Screen Printing**. Cincinnati, Signs of the Times Publishing Co., 1962.
Kulp, George, **Teaching the Designing of Decals and Hand Painted China: Lustre Colors and Effects, Gold Work, Lost Arts**. Florida, Atlantic Printers and Lithographers, Inc., 1970.
Rhodes, Daniel, **Clay and Glazes for the Potter**. New York, Chilton Book Company, 1957.
*Long, Lois Culver, **Ceramic Decorations**. Indianapolis, Indiana, The American Art Clay Corp., 1958.
*Standard Ceramic Supply Company 1976 Catalog: Clays, Chemicals, Kilns, and Tools**. Division of Chem-Clay Corporation, P.O. Box 4435, Pittsburgh, Pennsylvania 15205.

PHOTO PRESENTATION ON FABRIC:
TECHNIQUE AND MATERIALS
Kosloff, Albert, **Textile Screen Printing**. Cincinnati, Signs of the Times Publishing Co., 1966.
Valentino, Richard, and Mufson, Phyllis, **Fabric Printing: Screen Method**. San Francisco, Bay Books, 1975.
***Naz-Dar's Catalog No. 34:** Screen Printing Inks, Equipment and Supplies*. Naz-Dar Co., 1087 North North Branch Street, Chicago, Illinois 60622.

PHOTOGRAPHIC EMULSIONS
Wall, E.J., **Photographic Emulsions: Their Preparation and Coating on Glass, Celluloid and Paper, Experimentally on the Large Scale**. Boston, American Photographic Publishing Co., 1929.
Zucker, Harvey, "*Print Your Wagon*," **35mm Photography**, Summer 1973, pp. 28-31.
*Eastman Kodak Publications:
AJ-5 **Photographic Sensitizer for Cloth and Paper.**
AJ-12 **Making a Photographic Emulsion.**
Write: Eastman Kodak Company, Dept. 454, Rochester, New York 14650.
***New Horizons in Photography**. Rockland Colloid Corp., 599 River Road, Piermont, New York 10968. A catalog of Rockland products.

NON-SILVER PROCESSES AND RELATED
BACKGROUND IN EARLY SILVER ALTERNATIVES
Barna, John, **The Gum/Oil Print, Experimentation**. Portland, Oregon. Unpublished graduate paper.
Brown, George E., **Ferric and Heliographic Processes: A Handbook for Photographers, Draughtsmen and**

Fan Printers. London, DawBorn & Ward, Ltd., 1917.

Bookstaber, Dennis, "The Ambrotype," **Camera 35**, July 1975, pp. 34-35.

Daguerre, Louis, **Historical and Descriptive Account of the Daguerreotype**. New York, Winter House Ltd., 1971. A reprint of the 1849 edition.

Derkacy, D., and Jones, D., "The Carbro Process," **Petersen's PhotoGraphic Magazine**, July 1975, pp. 71-74.

Estabrooke, Edward, M., **The Ferrotype and How to Make It**. Hastings-on-Hudson, New York, Morgan & Morgan, 1972. A reprint of the 1872 edition. All about tintypes.

Fraprie, Frank R. & Woodbury, Water E., **Photographic Amusements**. Boston, Mass. American Photographic Publishing Co., 1931.

Gassan, Arnold, **Handbook for Contemporary Photography**. Athens, Ohio, Handbook Co., 1974. Good section on platinum printing technique, photogravure, and collotype printing.

Green, Dr. Robert, "Tri-Color Carbro," **Petersen's PhotoGraphic Magazine**, May 1977, pp. 78-87.

Hunt, Robert, **History of Photography: The Coloring Matter of Flowers**. pp. 163-192. Annual Report of The Smithsonian Institution. Washington D.C., Government Printing Office, 1906.

Kernan, Sean, "An Old-Time Technique: Palladium," **35mm Photography**, spring 1977, pp. 36-40, 111-112.

Kosar, Jaromir, **Light-Sensitive Systems: Chemistry and Aplication of Nonsilver Halide Photographic Processes**. New York, John Wiley & Sons, 1965.

Jones, Darryl, "Carbonated Prints," **35mm Photography**, Summer 1976, 72-81, 107-111. Good introduction to carbro and carbon printing technique.

Marino, T.J., **Pictures Without a Camera**. New York, Sterling Publishing Co., 1975.

Marton, A.M., **A New Treatise on the Modern Methods of Carbon Printing**. Bloomington, Illinois, Photograph Ptg. & Sta. Co., 1905. Explains how to make carbon tissue.

Nonsilver Printing Processes: Four Selections, 1886-1927, Peter Bunnel, ed. New York, Arno Press, 1973.

Snelling, Henry H., **The History and Practice of the Art of Photography**. Hastings-on-Hudson, New York, Morgan & Morgan, 1970. A reprint of the 1849 edition.

Tennant and Ward. **The Photo Miniature**. No. 185, Vol. 16, January 1922. New York. "Kallitype and Allied Processes: A Collection of Practical Methods for the Preparation of Photographic Printing Paper With the Iron Salts. Kallitype, Revised to Date, Iron-Silver, Uranium, Copper and Platinum Papers; With Formulas and Working Instructions."

Towler, J., **The Silver Sunbeam**. Hastings-on-Hudson, New York, Morgan & Morgan, 1969. A reprint of the 1864 edition.

Wall, E.J. & Jordan, Franklin I., **Photographic Facts and Formulas**. Boston, Mass. American Photographic Publishing Co., 1947.

Wolf, Linda, "How to Make Calotypes," **Petersen's PhotoGraphic Magazine**, May 1977, pp. 96-103. (mistitled article: Kallitypes not Calotypes)

PHOTO SILKSCREEN TECHNIQUE

Fossett, R.O., **Screen Printing Photographic Techniques**. Cincinnati, Signs of the Times Publishing Co., 1973.

Hiett, Harry L., **57 How-to-Do-It-Charts on Materials, Equipment, Techniques for Screen Printing**. Cincinnati, Signs of the Times Publishing Co., 1959.

Kosloff, Albert, **Photographic Screen Printing**. Cincinnati, Signs of the Times Publishing Co., 1972.

Kosloff, Albert, **Screen Printing Techniques**. Cincinnati, Signs of the Times Publishing Co., 1972.

*Atlas Catalog: **Inks, Supplies, and Equipment**. Atlas Screen Printing Supplies, 1733 Milwaukee Ave., Chicago, Illinois 60647.

*Stencil Fabrics for Screen Printing**. TETKO, Inc., 420 Saw Mill River Rd., Elmsford, New York 10523.

TRANSPARENCY MAKING

Curwen, Harold, **Processes of Graphic Reproduction in Printing**. London, Faber & Faber, Ltd., 1966.

Gould, John, "A Practical Guide to Posterization," **Camera 35**, July 1975, pp. 36-40.

Holter, Patra, **Photography Without a Camera**, New York, Van Nostrand Reinhold Company, 1972.

Lundquist, Par, **Photographics: Line and Contrast Methods**. New York, Van Nostrand Reinhold Company, 1972.

Routh, Robert D., **Photographics**. Los Angeles, Petersen Publishing Co., 1976.

COLOR SEPARATION:
BASIC CONCEPTS AND TECHNIQUE

Color. New York, Time-Life Books, 1970.

Henney, Keith, **Color Photography for the Amateur**, New York, McGraw-Hill Book Company, 1948.

Mante, Harold, **Color Design in Photography**. New York, Van Nostrand Reinhold Company, 1972.

Spencer, D.R., **Colour Photography in Practice**. London, Focal Press, Ltd., 1969.

Wall, E.J., **The History of Three-Color Photography**. Boston, American Photographic Publishing Co., 1925.

*Eastman Kodak Publications:

E-47 **For Newspaper Color Reproduction: Separation Prints From Color Negatives.**

E-48 **Three-Color Separation Prints From Color Transparencies for Newspaper ROB Reproduction.**

E-80 **Kodak Dye Transfer Process.**

Q-7 **Basic Color for the Graphic Arts.**

Q-114 **Kodak Direct-Screen Color-Separation Method.**

Q-121 **Kodak Contact Direct-Screen Color-Separation Method.**

Write: Eastman Kodak Company, Department 454, Rochester, New York 14650.

HAND COLORING PHOTOS
Baker, Gla, **How to Paint Photographs for Profit**. New York, Carlton Press, 1962.
Jerome, A., **Crayon Portraiture**. New York, New York, Baker & Taylor Co., 1892.
Tobias, J. Carroll, **The Art of Coloring Photographic Prints**. Boston, American Photographic Publishing Co., 1934. Formulas for sensitizing a variety of surfaces and techniques for hand coloring.
Wall, Alfred H., **A Manual of Artistic Colouring, As Applied to Photographs**. New York, Arno Press, 1973. Facsimile of an old original edition.
Walley, Charles. **Colouring, Tinting and Toning Photographs**. London, The Foundation Press, 1956.

COMMUNICATION EQUIPMENT, MATERIALS, AND CONTEMPORARY IMAGING SYSTEMS
Caulfield, H.F., and Sun Lu, **The Applications of Holography**. New York, John Wiley & Sons, 1970.
Clark. H.E., and Dessauer, J.H., **Xerography and Related Processes**. London, Focal Press, 1965.
Cook, W.A., **Electrostatics in Reprography**. London, Focal Press, 1970.
Dinaburg, M.S., **Photosensitive Diazo Compounds**. London, Focal Press, 1967.
Golwyn, Craig. *"This Issue: Electrostatics, State of the Science, State of the Art, State of Their Union,"* **Yony**, Vol. 1, No. 4, May 1975. **Yony** is the official publication of the Generative Systems Workshop at the School of The Art Institute of Chicago.
"Inventions: Solid Photography," **Newsweek**, July 12, 1976, p. 56.
Kripper, Stanley, and Rubin, Daniel, **The Kirlian Aura ... Photographing the Galaxies of Life**. New York, Anchor Press, 1974.
Land-Weber, Ellen, *"3M Montage to Create Dreamlike Images Using Modern Technology,"* **Petersen's PhotoGraphic Magazine**, May 1976, pp. 72-76.

Lehmann, Matt, **Holography: Technique and Practise**. London, Focal Press, 1970.
Scaffert, R.M., **Electrophotography**. London, Focal Press, 1965.
Williams, Alice, *"Jill Lynne—Collage to Create— 'Machine Art' With Xerox and 3M Color Copies,'* **Popular Photography**, March 1977, pp. 88-91, 174, 179.
*Everything You Wanted to Know About Your Own Photo Poster Blow-Up Business But Weren't Afraid to Ask Itek**. Itek Business Products, Market Development Publications, Rochester, New York 14603.
*"How to" Guide for Design Graphics: 3M Color-Key Contact Imaging Material**. Printing Products Division, 3M Center, St. Paul, Minnesota 55101. Booklet.

LITHOGRAPHY
Antreasian, Garo Z., and Adams, Clinton, **The Tamarind Book of Lithography: Art and Techniques**. New York, Harry N. Abrams, 1971. Complete book on lithography.
Banister, Manly, **Lithographic Prints From Stone and Plate**. New York, Sterling Publishing Co., 1972.
Knigin, Michael, and Zimiles, Murray, **The Technique of Fine Art Lithography**. New York, Van Nostrand Reinhold Company, 1970.

UNCOMMON PHOTO DISPLAY TECHNIQUES
*Ademco Dry Mounting/Laminating Equipment and Heal Seal Films**. Photo Technical Products Group, 623 Stewart Ave., Garden City, New York 11530. Literature and supplies.
*Decal'comania, The Art of Making Decals**. Sangray Corporation, P.O. Box 2388, Pueblo, Colorado 81004. 1975 booklet on technique.
*Eastman Kodak Publication G-12 **Making and Mounting Big Black-and-White Enlargements and Photo Murals**. Rochester, New York 14650.
*Quebradra Découpage**. McDonald Photo Products, Inc., 2522 Butler, Dallas, Texas 75235.